Spatial Evolutionary Modeling

SPATIAL INFORMATION SYSTEMS

General Editors

M.F. Goodchild

P.A. Burrough

R. McDonnell

P. Switzer

SPATIAL EVOLUTIONARY MODELING

Roman Krzanowski
Jonathan Raper

OXFORD
UNIVERSITY PRESS

2001

OXFORD
UNIVERSITY PRESS

Oxford New York
Athens Auckland Bangkok Bogotá Buenos Aires Calcutta
Cape Town Chennai Dar es Salaam Delhi Florence Hong Kong Istanbul
Karachi Kuala Lumpur Madrid Melbourne Mexico City Mumbai
Nairobi Paris São Paulo Shanghai Singapore Taipei Tokyo Toronto Warsaw

and associated companies in
Berlin Ibadan

Published by Oxford University Press, Inc.
198 Madison Avenue, New York, New York 10016

Library of Congress Cataloging-in-Publication Data
Krzanowski, R. M. (Roman M.)
Spatial evolutionary modeling / R.M. Krzanowski and J. Raper.
p. cm. – (Spatial information systems)
Includes bibliographical references and index.
ISBN 0-19-513568-7
1. Geographic information systems. 2. Genetic algorithms.
3. Mathematical models.
I. Raper, Jonathan. II. Title. III. Series.
G70.212.K79 2000
910′.285—dc21 00-037348

9 8 7 6 5 4 3 2 1

Printed in the United States of America
on acid-free paper

This book is dedicated
to Professor Stan Openshaw,
who pioneered geocomputational studies and
laid the foundations for this research

Foreword

MICHAEL F. GOODCHILD

Stan Openshaw, Professor of Geography at Leeds University, has built a reputation as a staunch advocate of computational methods for solving spatial problems. A decade ago he developed the first geographical analysis machine (GAM) to search through many alternative descriptions of patterns of human spatial interaction for new models. Other GAMs were built to search for anomalies in patterns of rare diseases such as leukemia. Stan was the foremost originator of a new field of geocomputation that promotes the use of computing to address and model complex geographic problems: there is now a center for geocomputation at Leeds, and a successful international conference series. It seems very appropriate, then, that the editors of this book have chosen to dedicate it to Stan, in recognition of his unique contribution to the invention of new computer-based methods for solving spatial problems, and of the exciting potential role of evolutionary modeling.

Evolution has populated the world with large numbers of highly successful organisms, each almost perfectly adapted to making use of the resources of its environment, and to resisting that environment's physical and biological hazards. The actual process of evolution is relatively simple, requiring only three mechanisms: one to produce significant variations in existing individuals in the next generation; one to ensure that only the most successful individuals of any generation survive; and one to preserve successful variations in future generations of individuals. If the environment is itself changing, then the process will work only if changes occur slowly in comparison to the lifetime of individuals; otherwise, it will not be possible for individuals to adapt successfully to environmental change.

Although the mechanisms are simple, with enough individuals and enough time the process will necessarily result in individuals that are almost optimally adapted to their environment. Of course the sheer scale of biological evolution is simply stunning: some 4 billion years, and numbers of individuals that can range into the trillions. Although the human race has had a mere 3 million years to evolve from its nonhuman ancestors, that is enough time for some 200,000 generations, and with 6 billion individuals now occupying the planet the potential for evolutionary adaptation is ample (although humans may now be changing their environment more rapidly than is prudent).

Many practical problems involve the search for optima against defined criteria. We humans are constantly solving such problems, as we pick routes to travel between places that minimize distance or travel time; as we design houses that provide an optimum level of comfort; or as we attempt to maximize the income we receive from investments. Many of these problems involve vast numbers of potential solutions, and it would be impossible for anyone to evaluate them all; instead, methods must be devised for solving them that cleverly exploit the structure of the problem. In the past, mathematics provided the basis for clever solutions, but only in certain circumstances; many classes of problems proved too difficult to analyze using mathematical tools. For example, the problem of finding the shortest tour through a known set of destinations—the traveling salesman problem—is known to be impossible to reduce using mathematical analysis, and consequently is very hard to solve. In principle, there are $n!/2$ possible tours through n destinations, and although some possible tours can be rejected out of hand, the number of reasonable alternatives for even a modest value of n is massive.

Computing power has grown enormously since the 1960s, enabling many new kinds of analysis. Evolutionary modeling exploits this increase in power, by emulating the behavior of a large population of individuals in biological systems as they go about their processes of reproduction, genetic diversification, and natural selection, with the implicit objective of optimizing themselves against the conditions that prevail in their environment. Of course it is still not possible to model the behavior of all of the billions of individuals of a species, but it is possible to reduce greatly the time taken for individuals to reach breeding age and reproduce, allowing these methods to produce useful results even with comparatively small populations.

This book is about the application of these methods of evolutionary modeling to spatial problems, or problems that involve searching for optima in geographic space. Geographic spaces tend to be enormously complex, requiring massive investment in their description and representation. For example, given a hard drive of 10 gigabytes it is possible to allocate only two words of plain language to the description of every square kilometer of the Earth's surface; and more detailed descriptions of much smaller areas typically run to terabytes. Many of the entities we design on the Earth's surface—buildings, cities, agricultural systems—are also complex. So spatial problems are particularly suited to evolutionary approaches that can harness the power of modern computers to analyze very large numbers of complex options.

The authors of this book have chosen a novel but intriguing approach, by beginning with a systematic and formal introduction to the topic, and then by inviting other contributors to write about applications. The result is a very useful compendium that will be an excellent text for specialist courses, and an important reference for researchers and users. Evolutionary modeling is still relatively unknown in the context of geographic information systems, a situation that is badly in need of correction if these systems are to live up to their claim to be indispensable for spatial decision support. This book will do much to correct that, and should encourage the builders of GISs either to offer evolutionary modeling directly, or to support its easy integration.

Preface

Motivations, Content, and Approach

Evolutionary models have been explored in scientific research for the past two decades. Yet, despite their conceptual attractiveness and versatility they remain largely unfamiliar to GIS researchers. This apparent lack of interest in evolutionary models among geographic information system (GIS) users may be due to the limited availability of the existing body of evolutionary research, largely confined to publications in the proceedings of specialized conferences. In addition, the esoteric language of this research literature may have contributed to an inadequate understanding of these modeling methods.

In this book we advocate the view that evolutionary-based modeling should be an integral part of a GIS analytical "tool box." Moreover, we believe that by overlooking the value of the evolutionary paradigm, the GIS field has been deprived of a powerful conceptual framework and a modeling technique which, when applied to spatial modeling, may offer important new insights into the nature of spatial models and help to develop solutions to problems earlier deemed overly complex and intractable.

This book was written to stimulate a wider interest of GIS researchers in applications of evolutionary modeling techniques for modeling of spatial phenomena. We envisioned it as a guide to evolutionary modeling. This book introduces the basic concepts of the evolutionary modeling methods, detailing the working principles of evolutionary models. It also presents several successful applications of evolutionary models of spatial phenomena. The terminology of evolutionary algorithms is thoroughly explained to ensure that the reader will not be intimidated by unfamiliar jargon.

We hope that this book will be of interest not only to GIS researchers but also to all those who are, in their professional activities, confronted by spatial problems. The applicability of evolutionary models to problems with a strong spatial component, such as bin packing, VLSI design, network design, and palette loading, has already been explored by researchers unrelated to the GIS field.

Intended Audience

The intended audience of this book includes GIS practitioners and researchers involved in data mining and modeling of spatial phenomena, senior and graduate students involved in advanced research on spatial data modeling, and members of the research community dealing with spatial problems in a wider, non-GIS-related, sense.

This book may also serve as a primary textbook for college-level courses in GIS, advanced spatial modeling method, and information engineering, or as a supplementary textbook for computer science courses.

Outline of the Book

The book is divided into three parts. Part I introduces the concept of evolution and evolutionary algorithms and covers the concepts, notation, and terminology that are needed to understand the later sections. It also provides a necessary background for a more comprehensive understanding of current research on evolutionary modeling. Furthermore, part I establishes the context in which evolutionary models and modeling are viewed in relation to more traditional information processing and modeling methods. We believe that there is a need to set a proper context for research and discussion of evolutionary modeling. As with every new method, there has been a lot of hype associated with evolutionary modeling. With it, came many expectations, frequently unfounded or misguided. Such an attitude may do more harm than good to a field and therefore we would like to correct some common misconceptions and misunderstandings about evolutionary modeling before they become established. In the early 1960s we saw an initial excitement and later disillusionment with artificial intelligence and its predictions of computers exceeding humans in every intelligent endeavor. Later, we witnessed the same disillusionment with neural networks. In the hope of sparing evolutionary methods a similar fate, in part I we shall clearly state what evolutionary algorithms can do and what they cannot. Although this approach may pour cold water on some fans of this new technology and turn off some enthusiasts, we believe that, in the long run, it will lead to more successful and robust applications of these algorithms.

The main focus of part II is the presentation of spatial evolutionary algorithms—the class of evolutionary algorithms designed to process spatial information. These algorithms are distinguished by the specific composition of their genetic material, designed to represent spatial information, and by the set of spatial evolutionary operators, designed to process spatial information. Both aspects of these algorithms (the more abstract and the algorithmic) are presented in detail. Further, part II contains a concise discussion of modeling in GIS that provides a context for the entire book. Part II concludes with a review of future research directions and it addresses some of as-yet-unanswered questions about applications of evolutionary models of spatial problems.

Finally, part III provides selected illustrations of the application of evolutionary algorithms to spatial problems by selected expert contributors. The presented examples range from an algorithm for the design of air space partitioning to a spatial learning algorithm. These studies offer a representative overview of current research into genetic modeling in GISs. The authors invited to submit manuscripts for inclusion in this part of the book have established themselves as the leaders in scientific application of evolutionary modeling of spatial problems.

Acknowledgments

Books are never created in vacuum. With every book there are people who helped along the way in many, sometimes unexpected ways. We would like to acknowledge them here.

Roman thanks Prof. S. Openshaw for his initial enthusiasm, which was critical in starting this project, the School of Geography at Birkbeck College for supporting the initial work on the idea of spatial evolutionary algorithms for four years, Dr. Randall Krausher for his help in clarifying some of the complexities of style and grammar, Bart Burns for his encouragement, long discussions and comments on the book, Dr. Aaron Dagen for comments on the text and for keeping the proper perspective on life, Ms. Joyce Berry of Oxford University Press for the patience and e-mails that kept this project alive, Dr. Eliza Krzanowska for merciless hunting of logical and other errors through the pages of this book, and Jacob Krzanowski for just being there. Very special thanks are also due to Angela Guimaraes Pereira who redrew figures in part II, making pure engineering drawings into works of art. Special thanks go to Bell Atlantic for supporting this project. Jonathan thanks the Department of Information Science at City University for the time to devote to this project.

We both thank Prof. Michael Goodchild for his Foreword, and the contributors who have added immeasurably to this book. We also thank Morgan Kaufmann Publishers for permission to use the GA Java code.

The authors would also like to acknowledge very helpful comments on the initial draft of the book provided by Prof. Marc P. Armstrong, University of Iowa. And last but not least, thanks to Sue Nicholls, Keyword Publishing Services Ltd., for detailed editing of this book.

February 2000

Roman Krzanowski
Jonathan Raper

Contents

Contributors

Christopher Brooks	University of London, London, U.K.
Mark de Berg	Utrecht University, Department of Computer Science, Padualaan 14, De Uithof, 3584 CH Utrecht, The Netherlands
Daniel Delahaye	CMAPX: Applied Mathematic Research Center (Ecole Polytechnique) LOG: Global Optimization Laboratory (Air Navigation Research Center), France delahaye@recherche.enac.fr
Catherine Dibble	University of Maryland, Maryland, U.S.A. cdibble@geog.umd.edu
Ângela Guimarães Pereira	New University of Lisbon – College of Science and Technology, Quinta da Torre, 2825 Monte da Caparica, Portugal angela.pereira@jrc.it
Roman Krzanowski	Bell Atlantic, White Plains, New York, U.S.A. krzanow@cloud9.com
Jonathan Raper	City University, London, U.K. raper@soi.city.ac.uk

Dirk Thierens Utrecht University, Department of Computer Science, Padualaan 14, De Uithof, 3584 CH Utrecht, The Netherlands

Steven van Dijk Utrecht University, Department of Computer Science, Padualaan 14, De Uithof, 3584 CH Utrecht, The Netherlands

PART I

EVOLUTIONARY ALGORITHMS

An Introduction

1

Concepts of Evolutionary Modeling and Evolutionary Algorithms

This book is about evolutionary algorithms as applied to spatial and geographic phenomena.[1] Why are we writing this book? Do these new algorithms deliver solutions to our modeling and data analysis problems that conventional methods cannot handle? Or will they just fade away, as have so many other "new" ideas from the past, some eventually finding their way into a museum of computer and conceptual contraptions?

It is not our purpose here to attempt to present answers to all of these questions. This is not because we lack expertise but because we do not yet know the answers. However, what we can do, and what we intend to do in this book, is to offer the reader a proper perspective within which to look at evolutionary models. For some, this perspective may prove disappointing, as we will not to solve all known information modeling ills. However, our perspective *will* help the reader to understand evolutionary computer methods and related concepts, and to use them in appropriate applications and models. In other words, we guarantee that the reader will not be disappointed in coming to a clear understanding as to what *evolutionary modeling is all about*. Admittedly, other books already serve this purpose. What we offer here, which we feel is unique, is a perspective on the new area of the application of evolutionary models—the area of spatial and geographic phenomena.

We shall begin with a broad introduction to models and modeling. This introduction will go well beyond the scope of computer science and geographic information systems (GISs) and will touch upon wider philosophical issues. We believe that modeling is a serious undertaking and it may have serious consequences for the modeler, the modeling subject, and even the lay public. In this introductory chapter, our objective is to assist the reader in coming to understand the modeling business as we see it (and would like others to see it).

Quest for Models

Modeling is as old as human civilization. Models shape what we are, how we define our reality, and what forms our thoughts may take.[2] To paraphrase a well-known saying:

> Tell me what your models are and I will tell you who you are.

Ancient cultures built models of the universe to predict their fate in the face of big unknowns and to explain incomprehensible phenomena. As primitive and naive as these models may now seem to us, each fulfilled a specific role within its particular scientific, social, and cultural context. Societies were shaped by them, as were the individual human lives. As the sophistication of the human race grew, and we accumulated experience in modeling, old models and social structures changed, but as they changed, we changed our concepts of society, ourselves, and the universe. We began to model flood cycles, lunar and solar eclipses, and phases of the moon. The ancient Greeks modeled with an accuracy that is comparable to the precision of contemporary modeling methods. We started to record, on media much more durable and shareable than human memory, our thoughts, our environment, our cities, and our travels. We also created a powerful, all encompassing logical model of reality—geometry.

So simple and yet so beautiful, geometry seemed to be a fundamental, or even *the* fundamental, language. With 10 basic principles, Euclid argued that we could model everything we needed to know about the shape of the universe. Euclid's geometric model of the universe was compact, elegant and, most of all, it worked. With it, we could build a straight road, partition land, and systematically assess (and collect) taxes. Geometric models seemed to be effective in solving practically every problem to which they were applied. Or so, for a long time, it seemed.

Even though there were occasional problems applying the geometric models in astronomy (some planets did not perform exactly as predicted), still, in most cases we were able to satisfy our need for order. Euclid's geometry was operationalized in Aristotle's dynamics and incorporated into the geocentric cosmology of the medieval Christian church in Europe.

Occasionally it became necessary, of course, to silence dissidents who persisted in pointing out the deficiencies of those early models and to dismiss significant departures from these models as products of witchcraft, "errors in observations," inaccuracies in sampling or, at worst, as merely insignificant deviations from the perfect order of geometric relations.

The dominance of this theological world view continued for hundreds of years. With geometry at our side, we could model the immediate aspects of our living space. The rest, we believed, was part of the Supreme Plan of God and we should not and could not inquire into its meaning and purpose.[3]

Then, at first quietly, the revolution began. In a small town in the eastern reaches of the Holy Empire a monk by the name of Nicholas Copernicus had been given the sole responsibility of rewriting the calendar. As he carried out that duty he came to understand that some of our models of the universe did not hold up well to close scrutiny: planets did not show up in the places and at the times expected, and new objects unexpectedly showed up in the sky. Being more of a pragmatist than a religious person, Copernicus proposed a new, simpler picture of Heaven that could explain some of these disturbing "anomalies." He then supported his position with careful scientific observations.[4]

Copernicus did not know that what he discovered about the Aristotelian models of dynamics (that they do not always fit observations!) would provide a template for centuries of human scientific endeavor to come. Copernicus did not know that the "new model" of the world that he had spawned would change forever what and how we think about our universe, nature, ourselves, and the world in which we live and die. He did not know that because of his insistence on correcting errors in models, symbols of authority would fall, wars would be fought, and new countries would be born (Davis, 1996). Such is the power of models.[5]

As we step back and survey the history of our human civilization, it becomes obvious that our ability to "model" (that is, to represent and interpret the world around us) is quintessential to who we are and how we come to define ourselves. In fact, it seems that, as we have created increasingly sophisticated models, we have opened up new realms of knowledge and new possibilities. It is also evident that advances in science, which frequently preceded changes in our society, were most often triggered by people who were not completely satisfied with dominant paradigms and concepts. They were people who felt deeply constrained by limitations of the existing models; individuals whose need for order and parsimony was offended by the deficiencies of existing systems to accurately represent reality, or to reliably predict its course.

As our discussion brings us into the modern age, it becomes increasingly obvious that the range and sophistication of models at our disposal is unprecedented in the history of science. It seems that we can model just about anything in the natural world. Not only can we model but also we can do that with the speed and accuracy that were never possible before. Yet, even today, as in ancient times, when we take a closer look at our models, errors, omissions, and simplifications often creep in and cannot be ignored. It is a curious paradox that the more thoroughly we describe what exactly it is that we want to model through our increasingly precise analytical methods, the more we come to realize that the phenomena we are trying to model pose more and more profound problems! Our dilemma approaches absurdity as it becomes apparent that our models cannot even handle all of the information that is at our disposal. It seems that we have amassed so much complex data that we have surpassed a point at which our models can function effectively. To some, it might even appear that data we have gathered is ultimately self-defeating![6]

Yet our need for better, more efficient, and more accurate models is greater than ever. The quest continues today as it has for centuries. As we are faced with new challenges, new questions to answer, new problems to solve, and new information to comprehend, we must also seek new models and new ways of expressing the same unknown universe that our ancestors faced. Our book is about exactly that: the presentation of a new modeling paradigm and new ways to represent it (Raper, 2000).

Evolutionary Models of Everything

Although the approaches and techniques described in this book do not bear comparison to Copernican revolution, nonetheless they have great potential as problem-solving tools in computation.

Even though evolutionary models may never inspire great shifts in human history, we are convinced that they will change the way we model. Through the use of evolutionary models we can explain hidden inner workings of our economies and make more comprehensible many complex phenomena of society and science, including such diverse matters as economics, the ecosystem, the human autoimmune system, highway traffic, or the organizational structure of an anthill (Holland, 1995; Gell-Mann, 1988). The paradigm of evolutionary models can change the way we view, explain, and model the most fundamental mechanisms underlying our reality.

The new approach to modeling (some people refer to it as a "paradigm") discussed in this book is an attempt to mimic the most ancient and universal process known to us, a mechanism that has shaped every living thing in our world. That process is known as evolution. We believe, and will show it experimentally, that our evolutionary model can represent complex phenomena in our world much better than existing models. But there is more to evolutionary models than merely improved modeling capability. The evolutionary model paradigm introduces a radically new approach to modeling and representation and it is likely to fundamentally change the way we model.

The procedure followed when using the "old" modeling paradigm required that we first try to guess the structure of the problem, then express it in some form of symbolism, and finally manipulate these symbols using some rules.[7] In evolutionary modeling, we actually do the very opposite: we make our representation of the problem model itself. We shall try to elucidate these ideas in the following sections.

Evolution was recognized and described by Charles Darwin more than a century ago. The discovery of evolution or, more accurately, the principles of evolution, was in some sense like the "discovery" of America 500 years ago: evolution was always "there," we just did not see it. Evolution was so ingrained in our everyday lives that nobody ever bothered to explore its potential in computation.

The ground-breaking idea advanced by Darwin in his famous book on the origin of species[8] (the book in which the principles of evolution were explicitly formulated for the first time) was that the development of life on earth is due to a slow process of gradual change across generations. Evolution works in the following way. All organisms are born with inherent differences and no two organisms in a population of organisms are ever *exactly* the same. During the organism's lifetime, organisms that are better suited to particular conditions in the environment produce more offspring than the organisms that are not as well suited to that setting. Over the course of *many* generations, the better-suited organisms will eventually "take over," while others will gradually die out. Should conditions in the environment change, organisms with particular features that are beneficial in that new, changed environment will survive and dominate, while others will eventually vanish.

If it is so simple for the organisms to adapt to their environment, why is it that the dinosaurs could not survive? It would appear that they were unable to adjust (or so it seems to most scientists) to changes in their environment—either because they ran out of useful (adaptive) diversity in their species, or because the environmental changes were too rapid and they simply ran out of time. Evolution takes time. Lots of it.

Regardless of the biological tale to be told, the essence of the story is, as follows: under environmental pressure (e.g., predation from natural enemies) the characteristics of the population of individuals will undergo changes in the direction of the pressure, that is, in the direction facilitating the survival.

Admittedly, we have not nearly done justice to the true complexities of the theory of evolution. Dialogue regarding the purpose of evolution, its sources, and its role in development of life is far from over. Some of the issues that have been closely scrutinized and, re-evaluated are the fundamental concepts of fitness and its role in the selection and the definition of a basic evolving unit and levels of evolution. Readers interested in such questions are invited to explore books by Sober (1993) and by Lloyd (1993). Widespread interest in concepts of evolution was manifested through the acclaim achieved by several books on the topic, one being written, and promoted, by Dawkins (1976). Dawkins presented some very interesting insights into the nature of evolution, which have been widely read. However, it became evident that Dawkins' later claims of new insight into the nature and existence of God, purportedly derived from studies of genetics and evolution or about the social and political consequences of these new insights perhaps overextended the power of evolutionary concepts. In contrast, authors such as Maturana & Varela (1992) presented a more enlightened, though less popular (or popularized), view of the role of genetics in evolution and heredity. These authors challenged the "deification of a DNA role" in Dawkins' work, observing that the error in the interpretation of the role of DNA (and genes) in evolution "lay in confusing the essential participation with unique responsibility." Their view seems to succinctly make the point that genetics and evolution can fall victim to a common problem of oversimplification. A similar view on Dawkins' claims has been expressed by the philosopher-futurologist S. Lem (Swirski, 1997).

Our intention here was to quickly sketch, for the benefit of the reader, the general principle that underlies *adaptation* under changing conditions. What Darwin could not possibly foretell in his time was that his principle of adaptation (he called it *evolution*) would eventually be applied to explain not only fundamental processes governing evolution of organisms, but also the formation of our consciousness, our memory and cognition, the development of the brain, human autoimmune system, our language, our economy, our societies and our technology. In recent years, this basic principle of adaptation has become known as a fundamental mechanism of what we call the label *complex adaptive systems* (CAS) (Gell-Mann, 1988; Holland, 1995, 1998; Hofstader, 1979).

Evolutionary Algorithms

As we have already pointed out earlier in this chapter, traditional modeling methods are based on formal, symbolic models of the real world. Typically, such models employ mathematical expressions[9] based on "hard-to-satisfy" assumptions.[10] The assertion that traditional models do, in fact, reflect reality comes from the assumption that reality can be successfully modeled by the formal logic apparatus.[11]

The limitations of traditional models are not really surprising as models can only be as good as their assumptions (Simon, 1996). Models based on the assumption of continuity can model only continuous realities. Linear models can only see linear realities. Normal models can only see Gaussian "worlds"[12] and formal logic models can see only formally logical relationships. The acceptance of a particular modeling algorithm implies the acceptance of the certain formal model of the world with all its limitations. Yet, all these limited worlds fall short of reality. As Mitchell (1996) said in her book about models:

> More recently we even come to understanding some fundamental limits to our ability to predict (model). Over the eons we have developed increasingly complex means to control many aspects of our lives and our interactions with nature, and we have learned, often the hard way, other aspects are uncontrollable.

Mitchell's comments about the limitations of our models and complexities of real phenomena are echoed by yet another author, H. Poincaré (1952), who almost a century ago expressed the following opinion about the continuity assumption underlying most of the mathematical models: "It is enough to warn the reader that the real mathematical continuum is quite different from that of the physicist. . . ."

Upon sifting through the scientific literature one finds many more examples of cautionary statements pertaining to other fundamental assumptions. One such example relates to the normal distribution, which has been named a "Gaussian disease" and yet another to linearity, which has been termed the "curse of linearity" (Isaaks & Srivastava, 1990).

Today, we repeat the history of scientific advancement. As it was in the past, today's scientists increasingly realize that our models are not very good at representing the problems they claim to model. As before, seemingly "small errors" find their way into our predictions, thus making it very difficult to maintain the integrity of our models. The time is right for history to repeat itself yet again, and for yet another evolutionary step in our thinking and in our science.

Partially in response to the need for newer modeling paradigms and better, more accurate models, modeling methods mimicking natural evolution appeared (Holland, 1993, 1995).[13] These evolutionary modeling methods, also known as evolutionary algorithms, have been afforded considerable attention over the recent years due to their ability to master the type of complexities that are beyond the grasp of traditional models. Evolutionary methods have offered an appealing alternative to the logic-based modeling methods for certain kinds of problems that are difficult to represent or that do not easily conform to the constraints of existing models. The types of problems for which this new approach to modeling could be applied have been characterized by Banzhaf et al. (1998):

1. [problems] where the interrelationships among the relevant variables are poorly understood,
2. [problems] where finding the size and shape of the ultimate solution to the problem is a major part of the problem,
3. [problems] where conventional mathematical analysis does not, or cannot provide analytical solutions,
4. [problems] where an approximate solution is acceptable (or is the only result that is ever likely to be obtained),
5. [problems] where small improvements in performance are routinely measured and highly prized,
6. [problems] where there is a large amount of data, in computer readable form, that requires examination, classification, and integration (such as molecular biology for protein and DNA sequences, astronomical data, satellite observation data, financial data, marketing transaction data, or data on the world wide web).

For these and similar types of problems, we cannot simply design a solution algorithm because we do not know how to specify the solution (that is what the algorithm does). As odd and improbable as an idea of modeling such problems might appear, it is, in fact, exactly what evolutionary modeling methods are designed for. They solve these types of problems not by telling the algorithm how to solve the problem, but by allowing the algorithm to find the solution by itself.

Basic Concepts of Evolutionary Algorithms

Evolutionary algorithms came out of research into complex adaptive systems which, in terms of their structure, resemble closely natural evolution.[14] Evolutionary algorithms manipulate *individuals* (or data structures), which undergo a set of changes and transformations. Each individual contains information about its properties in a structure called a *gene*. Each individual is also assigned some metric that expresses its value (fitness), that is its role, in achieving the objectives of the algorithm. And, as with any algorithm, there are rules to halt its operation.

We may ask, exactly how closely do evolutionary algorithms resemble natural evolution? Speculating about this relationship in a context of genetic programs (GP)—"close cousins" of evolutionary algorithms[15]—Banzhaf et al. (1998) observed that:

> A GP algorithm was inspired by the theory of evolution . . . . No claim is made . . . that the GP algorithm duplicates biological evolution or is even closely modeled on it. At most we can say that GP algorithms have been loosely based on biological models of evolution and sexual reproduction.

Although this observation was made specifically about GPs, it applies to any form of evolutionary modeling. While the analogy to natural evolution is mostly conceptual, it is, no doubt, the main source of inspiration for all evolutionary models.

Evolutionary algorithms have developed a lexicon with highly specialized terms and definitions. The familiarity with this terminology is essential for understanding of the structure and operations of the algorithm.

In the "language" of evolutionary algorithms, the *individual* denotes an evolving unit of the evolutionary algorithm, which carries genetic material in one or more of its chromosomes. A *gene* is a section of a chromosome responsible for one feature of the individual.[16] A set[17] of individuals is called a *population*. *Fitness* is a measure of the "value" of an individual with respect to the objectives of the evolutionary algorithm. The *objective function* is a function mapping the fitness of individuals into the problem space. It is typically the same as the fitness function. The *generation* is one evolutionary cycle. The *parent population* is a set of individuals of the previous generation. The *offspring population* is a set of individuals of the next generation. *Evolutionary operators* are transformations carried out on individuals during evolution. The most important evolutionary operators are cross-over (reproduction), mutation, and selection. *Cross-over* is the process of exchange of genetic material between individuals. *Mutation* is a random change of the genetic material of the individual. *Selection* is the process of designating individuals from a population for cross-over. *Initialization of a population* is a process of generating an initial population of individuals. A *terminating function* is the function that defines the condition that stops evolution. A more formal definition of these concepts is presented later in this section.

A simple evolutionary algorithm may be presented as series of six steps:

step 1: initialize population of individuals
step 2: assign fitness to individuals
    step 3: select individuals for cross-over
    step 4: perform mating and cross-over
    step 5: assign fitness to new individuals
    step 6: if stopping criteria not met
            repeat from step three,
        otherwise STOP

We begin evolutionary modeling by designing the representation of the problem we want to model as a population of some individuals and by selecting appropriate operators. A very important task, at this stage, is to design a fitness operator. The fitness operator will calculate the fitness score of our individuals and will guide the evolution of our model. A poor fitness function will result in a poor model and a good fitness function will give us well performing model. Now we can generate a set of individuals that represent some solution of the modeled problem (Step 1). These individuals are generated without regard to the actual *objective* of the model, but they must be "located" within the domain of the model. Each individual is assigned a number—fitness—which expresses its value for the model (Step 2). The fitness value ranks individuals from best to worst.

The actual algorithm which calculates fitness values is dependent on the model. Out of initial population of individuals a number of individuals is selected using some selection method (Step 3). The selection method may select the upper 50 percent of individuals with highest fitness values or it may use some other selection mechanism to create the mix of individuals with different fitness values. The selected individuals are paired randomly in the process of *mating*. The two mating individuals exchange their genetic material in the operation of *cross-over* and create a new *individual*. The process of mating and cross-over continues for as long as the desired number of new individuals is created (Step 4). These individuals constitute a new population. Fitness is calculated for them (Step 5), and the process of selection–mating–cross-over is repeated until designated criteria for termination are met (Step 6). During evolution, that is the operation of the algorithm, other operators may intervene. For example, *mutation* causes random changes in chromosomes and *learning* changes the fitness of individuals. By the time the algorithm terminates, all the individuals are usually the same or, more precisely, they have the same (or very similar) fitness. These individuals represent the solution to the model.

Figures 1-1 through 1-3 provide a conceptual representation of the fundamental components and operations of evolutionary models. They may help the reader to visualize the elements of an evolutionary algorithm. Figure 1-1 presents the components of evolutionary algorithms—genes, chromosomes, individuals, and populations. Figures 1-2 and 1-3 present two fundamental evolutionary operators, *cross-over* and *mutation*.

Cross-over was earlier explained as a mechanism that combines the chromosomes of two individuals and creates a new individual whose chromosome(s) represent a certain combination of chromosomes of parent individuals. As

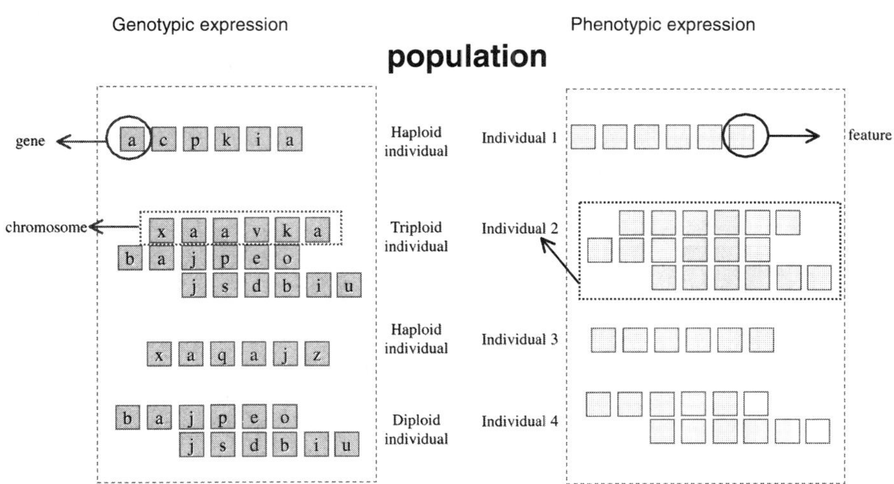

A population is composed of individuals. Each individual contains one or more chromosomes. Each chromosome in turn is composed of several genes. Genes are represented here as squares with letters and chromosomes as strings of squares (genes). For example we have individuals with one chromosome [acpkia], two chromosomes [bajpeo, jsdbiu], and three chromosomes [xaavka, bajpeo, jsdbiu]. Each chromosome has six genes. The chromosome [acpkia] has genes: a, c, p, k, i, a. A population has four individuals: two with one chromosome – haploid individual, one with two chromosomes – diploid individual, and one with three chromosomes – triploid individual.

Figure 1-1 Components of a genetic algorithm.

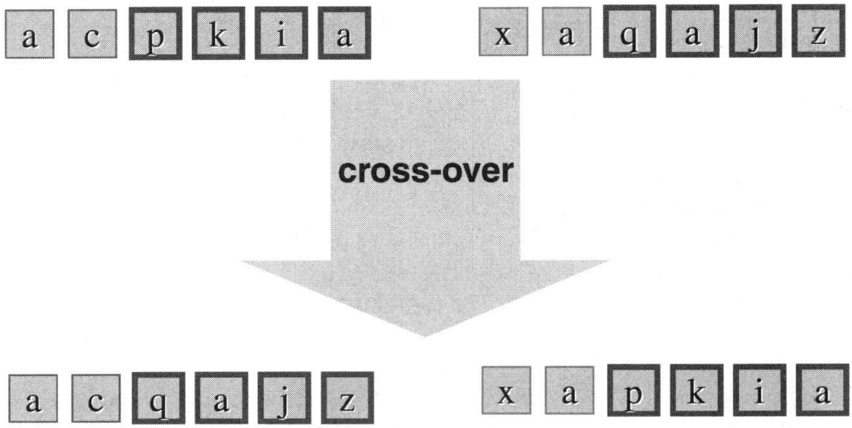

Cross-over enacts the exchange of segments (parts) of two chromosomes of different individuals. As seen here, during the cross-over four genes "pkia," from a chromosome on the left [acpkia], are exchanged with four genes "qajz," from a chromosome on the right [xaqajz], and new chromosomes [acqajz] and [xapkia] are created.

Figure 1-2 Cross-over operation between chromosomes of two individuals.

shown in figure 1-2, during cross-over chromosomes of mating individuals are divided and the separated parts of the chromosomes are exchanged. Depending on the type of cross-over, division of the chromosome may occur in one segment, in two segments, or in many segments.

Mutation is an operator that changes the value of a gene. Both the position of the mutated gene, as well as the change of the gene are random. As illustrated in figure 1-3, mutation affects one of the genes in the chromosomes. Mutation may be beneficial to the individual when it increases its fitness. But it may also have a detrimental effect on fitness.

There are many variations of each operator. Some of these variations will be discussed later in this chapter. However, before going further in our discussion of evolutionary models, a practical example may be useful in order to demonstrate basic concepts of evolutionary algorithms. This example is very simple but it nicely illustrates all basic principles of evolutionary algorithms discussed so far.

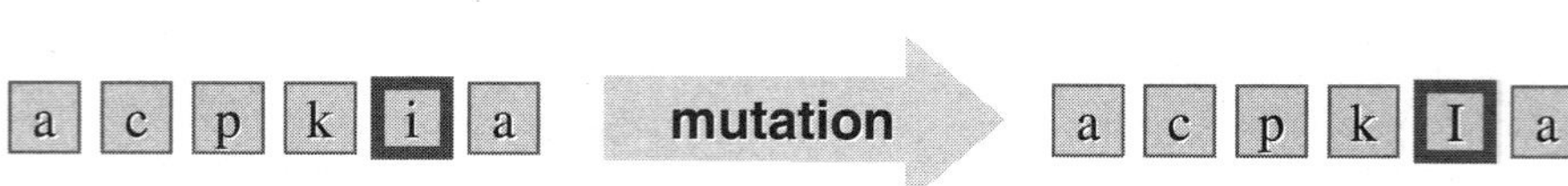

Mutation enacts the change of a gene in a chromosome. Here, one gene "i" in a chromosome [acpkia] is changed into a gene "I". The individual, the chromosome, and the gene that are mutated are selected randomly.

Figure 1-3 Mutation operation on a chromosome.

A Simple Example: $f = x^2$

In this example we want to find the minimum of a function $f = x^2$ using an evolutionary algorithm. We know a priori the solution to this problem. It is $x_{0.0} = 0.0$. We start our evolutionary modeling by defining the domain of the model and individuals that represent possible solutions to our model. The domain of the model is a closed interval $[-10.0, 10.0]$. Each individual in our model is represented by x and it may have any value from the interval $[-10.0, 10.0]$. The value of each individual is encoded in its chromosome as a real number. The chromosome is a data structure that holds the value assigned to the individual. If our evolutionary model were implemented as a simple computer program in C, a chromosome could be represented as a variable of a type float of 4 bytes length.

We use the following operators in our algorithm: *initialization, selection, mating and cross-over, mutation*, and *fitness*. The initialization operator is used to generate the initial solution to the model (the initial population of individuals—Step 1 in our algorithm). The initial population is created with a formula $rnd_{\langle 0,1\rangle} * 20.0 - 10.0$, where $rnd_{\langle 0,1\rangle}$ is a uniform random number in a $\langle 0.0, 1.0\rangle$ closed interval. The selection operator is Tournament 1-1, which works in the following way. Two individuals are randomly pooled from the population. Out of these two individuals, the one with the higher fitness is put into a new population and both individuals are returned to the original population. This process is repeated until the number of individuals in the new population equals the number of individuals in the original, parent population. From the new population two parent individuals are selected for reproduction (*cross-over*) using the mating operator. These individuals are crossed over, producing one individual that is placed in the offspring population. Parent individuals are returned to their population.

The process of mating and cross-over is repeated until the number of individuals in the offspring population equals the number of individuals in the parent population. The cross-over operator is implemented using the formula $x_l^j = (x_i^{j-1} + x_k^{j-1}) * 0.5$ which is an arithmetic average of chromosomes. In this formula, l is an index of a new individual, j is an index of a new generation, $j-1$ is an index of a parent generation, and i, k are indices of individuals in a population of a $j-1$ generation. Fitness of individuals is calculated as a distance (a square of the difference) of a chromosome from the optimal value of the function $f = x^2(x = 0.0)$ using the formula $f(x_i) = (x_i - x_{0.0})^2$. The mutation operator is implemented as a multiplication of a chromosome by a uniform random number $rnd_{\langle 0,1\rangle}$ using the formula $rnd_{\langle 0,1\rangle} * x_i = x_{iM}$. The mutation operator is activated at randomly selected generations when a random number pooled for each generation falls below some preset value. We call this preset value a mutation activation probability. This design of mutation assures that the number of times this operator is activated during the evolution is controlled. Individuals are pooled for a mutation from the selected generation randomly, with each individual having the same chance of being selected.

Table 1-1 shows how the initial population of five individuals is derived. The column "ID" contains the index of each individual.[18] The next column is a random number used to generate an individual in the column "x_i." The column "$x_i - x_{0,0}$"

Table 1-1 Initial population of five individuals for the genetic model of $f = x^2$

ID	$rnd_{\langle 0,1 \rangle}$	x_i	$x_i - x_{0.0}$	$f_i = (x_i - x_{0.0})^2$	$f_{in} = f_i / \Sigma f_i$	f_{in}^{-1}
1	0.257	−4.86	−4.86	23.61	0.146	6.84
2	0.877	7.54	7.54	56.85	0.352	2.84
3	0.754	5.08	5.08	25.80	0.160	6.25
4	0.149	−7.02	−7.02	49.28	0.305	3.27
5	0.619	2.38	2.38	5.66	0.035	28.57
			Σf_i	161.2		

is the difference between the individual and the optimal value of the function $f = x^2$. The column "$f_i = (x_i - x_{0.0})^2$" gives the fitness of an individual.

The last two columns in table 1-1 give the reader a taste of how genetic modeling is done in real life. The column $f_{in} = f_i / \Sigma f_i$ represents *normalized fitness* (fitness expressed as a fraction of the total fitness of all organisms in a population). This representation allows us to attribute probabilities to each of our fitness scores. As relative scores can be represented using a cumulative density function, each relative or normalized fitness score can be represented as a fraction of a total score. Relative fitness scores are used in certain selection operators (such as a roulette-wheel).

The column f_{in}^{-1} represents the inverse of a normalized fitness score. This form of fitness score makes it easy to observe that the best fitness is also the largest. As noted previously, in our simple example, it really does not matter which representation of the fitness score we use. Why do we square the fitness score? Although it is not very critical, we do it in order to avoid having negative numbers as fitness. It simplifies the comparison of fitness scores. In our example, if the fitness scores were not scored, two individuals with the same distance to the optimal value of the model would have different fitness scores (for example (−1.2) and (1.2)). By squaring the fitness function we avoid this situation.

A word of caution. In real-life applications of evolutionary models the value of the fitness function cannot be calculated as in our example, because the optimal value of the model is unknown. For if we *knew* it, what would be the purpose of the model? Thus, in these applications the fitness function is usually much more difficult to define.

Given the initial population with assigned fitness scores, it becomes possible to observe the first evolutionary cycle. This cycle is demonstrated in table 1-2. The

Table 1-2 A first evolutionary cycle of selection, cross-over, and fitness

Tournament 1-1	Selected Individual	Mating Cross-over	x_i^j	$f_i = (x_i - x_{0.0})^2$
5–2	5	5–4	−2.32	5.38
4–5	5	5–3	3.73	13.91
2–3	3	3–4	−0.97	0.94
5–1	5	3–5	3.73	10.17
3–4	4	4–5	−4.64	21.52

Table 1-3 The effect of mutation on an individual

x_i	f_i	$rnd_{\langle 0,1 \rangle}$	x_{iM}	f_i
5.08	28.80	0.01	0.05	0.002
5.08	28.80	0.98	4.97	24.78

column labeled *Tournament 1-1* shows the indices of individuals paired in the Tournament 1-1 selection. The next column shows selected individuals. The *mating and cross-over* column shows the individuals that have been selected for mating and reproduction. The next column shows the individual that was created in the cross-over of the individuals from the previous column. Finally, the last column shows its fitness.

Upon viewing the last column that represents fitness scores of individuals in the offspring population, the reader will notice that the fitness scores of individuals in this population are much lower than in the parent population (table 1-1). "Lower" means "better" in our example where individuals with lower fitness scores cluster closer to the optimal value of the model. The improvement of the fitness scores is exactly what the evolutionary cycle of selection and mating is all about.

The evolutionary process, described above, is repeated until fitness of all the individuals in the population has converged to the optimum of the modeled function (within the predefined error margin) or has become the same. This "predefined distance" is defined as a stopping criterion. In our example of the evolutionary model it may happen that they never reach the exact value of 0.00 as genetic algorithms often provide only suboptimal (or, "close-to-optimal") solutions.

To keep our example simple we did not demonstrate the effect of *mutation*. In evolution, mutation may have one of two effects on the population and individuals. It may *improve* fitness but it can also make it *worse*. In our example, the mutation operator is implemented in a way that will always result in an improvement of fitness scores of individuals. Table 1-3 demonstrates the effect of mutation on the fitness score of an individual. In table 1-3 column x_i shows an individual selected for mutation. The next column shows its fitness. The following column shows a random number used in the mutation. Column x_{iM} represents the mutated individual, while the next column shows its fitness.

We must now revisit our discussion about evolution and Darwin. In table 1-4 we provide a rough comparison of concepts of natural evolution and evolutionary models with elements of an evolutionary model of a function $f = x^2$. There is an obvious analogy.

Our example of a genetic algorithm was simple and easy to understand. It was also a type of example that anyone knowing anything about the evolutionary models such as "New Modeling Paradigm" would expect. After all, evolutionary methods are optimization algorithms and finding the minimum of $f = x^2$ seems like a perfect problem for them.

Table 1-4 Mapping of concepts between natural evolution, evolutionary algorithms, and an evolutionary algorithm model

Concepts of Natural Evolution	Concepts of Evolutionary Models	Elements of the Example Model
Population	A collection of strings or data objects	$x_i, j = 1, \ldots, n$
Individual(s)	Data objects, strings, arrays, lists, etc.	x_i
Chromosomes	As above	Real *4
Genes	Part of data objects or data objects themselves	Value of x_i
Selection	Operators	Tournament 1-1
Mutation		$rnd_{\langle 0,1\rangle} * x_i = x_{iM}$
Reproduction (cross-over)		$x_l^j = (x_i^{j-1} + x_k^{j-1}) * 0.5$
Fitness	Value (metric) assigned, associated with a data object	$f_i = (x_i - x_{0.0})^2$
Evolution	A cycle of the algorithm from initialization to stop	Algorithm

Holland Example

Our next example, taken from Holland's book *Hidden Order* (Holland, 1995), is more surprising. It is surprising because it demonstrates the use of an evolutionary model with the problem that one would not expect to have anything to do with evolution and evolutionary algorithms. However, after overcoming any initial skepticism, we will be able to see commonalties and parallels between our first example of the evolutionary algorithm that models $f = x^2$ and the model presented by Holland. We will also find it easier to grasp the real nature of evolutionary computation and what it may be used for.

Our new model is that of a frog trying to catch a fly. The frog sees an object—a prey, predator, or perhaps something that is neither of them. The frog has to make a decision about what to do based on what it sees. It may run away, pursue the object, turn its head, or extend its tongue. As simple as this decision may seem, in reality this frog makes a decision between staying alive and having a full stomach and being dead and in someone else's stomach. It is, in fact, a decision between life and death. How does this frog come to the most optimal decision? By experience and observation. The frog learns to associate the proper response {run away, pursue, turn head, extend tongue, stay still, jump, ...} with the proper stimuli {small object, big object, object with red legs, near, far, fast, slow, ...}. We may imagine the frog's decision system working as a set of IF-THEN rules. IF {small object, big object, object with red legs, near, far, fast, slow, ...} THEN {run away, pursue, turn head, extend tongue, stay still, jump, ...}.

We may also imagine that the young frog has initially all these rules mixed up like this:

IF {big} THEN {jump}
IF {small} THEN {runaway}
IF {near} THEN {run}
. . .
etc.

But every time the frog gets the rule right it gets its lunch. And every time it does not get the rule right, it stays hungry. In this way rules that promote its well-being get remembered better and used more often. Rules not-so-successful get forgotten or eliminated. With time, the frog gets the rules right—most of the time. By the way, frogs that do not learn the rules fast enough, die.

It should be obvious what is the underlying model of the frog's decision making. There is a set of rules that associate the stimuli and the responses. The rules are initially random. Every rule has the associated fitness. There is a process of selecting and combining the rules to get new rules. One may say that the rules evolve by trial-and-error. And there is a process of assessing whether the given combination of rules is more or less beneficial to the frog.

Indeed, in the frog's behavior we have all elements of an evolutionary model; the frog uses an evolutionary algorithm, of sorts, to get the rules right and survive.

Forms of Evolutionary Algorithms

The class of evolutionary algorithms includes a number of different methods that are variations of the basic algorithm. To this class belong evolutionary programs, evolutionary programming, evolutionary strategies, and genetic algorithms, along with their variations (canonical GA, hybrid genetic algorithms, messy GA, modified GA, and others). There is a lack of consensus regarding a taxonomy of evolutionary methods. Table 1-5 presents a high-level classification of evolutionary modeling techniques based on Michalewicz (1992, 1993), Back (1996), and Mitchell (1996). Although this classification encompasses most of the major varieties of evolutionary algorithms, it does not purport to present all of them. Such a task is probably impossible, since practically every researcher that has contributed a new operator to the basic structure of evolutionary algorithms added his/her own class. We shall discuss various forms of evolutionary algorithms later in this book.

Applications of Evolutionary Algorithms

The types of problems successfully modeled by evolutionary methods include problems that are poorly understood; that have complex search spaces (so-called "hard" problems) (Liepins & Hillard, 1989); or that are represented by

Table 1-5 Taxonomy of major classes of evolutionary algorithms

Method	Major Characteristics	References
Classical (or, canonical) genetic algorithms	A class of search algorithms using binary coding, fixed-length binary strings, selection, mutation, and cross-over	Holland (1993)
Evolutionary programming	A class of search algorithms for optimization of continuous parameters using mutation; a method similar to evolutionary programs.	Fogel (1991)
Evolutionary algorithms	A class of search algorithms based on the concept of evolution. This is an equivalent definition to the Michalewicz definition of evolutionary programs	Back (1996)
Evolutionary programs	See evolutionary algorithms	Michalewicz (1992)
Evolutionary strategies	A class of evolutionary algorithms using floating-point representation and mutation	Schwefel (1981)
Hybrid genetic algorithms/ Hybrid evolutionary algorithms	A class of search algorithms incorporating problem specific knowledge or non evolutionary operators	For example: Al-Attar (1994), Medsker (1995), Fleurent & Ferland (1994)
Genetic programming	A class of evolutionary algorithms designed to search for new computer programs	Koza (1991)

nondifferentiable functions or by functions with multiple local minima (Whitley, 1993). Some of these problems are listed below.

- Optimization problems, such as multimodal function optimization, image processing, optimization of the combinatorial problems, such as traveling salesman problem (TSP), bin packing, mechanical design.
- Automatic programming such as generation of computer programs for specific tasks.
- Machine learning for classification, prediction tasks, and data mining (Celko, 1993; Michalewicz, 1992).
- Scheduling and rule learning such as job shop scheduling, nuclear power plant design scheduling, airport traffic control (Delahaye et al., 1994; Davis, 1991; Nissen, 1993).
- Economic modeling including the development of bidding strategies, predictions of stock markets. Analysis of social systems including studies of emergence of social behavior in insect colonies, evolution of cooperation.
- Immune system modeling.
- Modeling of intrusion detection systems in computer networks.

An extensive list of references on the application of evolutionary algorithms is provided by Allander (1995a–h).

Evolutionary Algorithms as Search Algorithms

Evolutionary algorithms are search algorithms. A search in this context means a "search for the solution among a set of candidate solutions." This is in contrast with a "search for stored data" or a "search for paths," which are other commonly accepted meanings of search in computer science (Mitchell, 1996). A set of all the possible solutions constitutes a search space. In an evolutionary algorithm, points in the search space are represented by individuals. In the context of this book a search space can also be a physical two-dimensional space. Evolutionary search is a sequence of transformations of individuals, with each individual representing a candidate solution. Transformations increase the frequency of individuals located in more favorable areas of the search space and decrease the frequency of individuals located in less favorable areas of the space. The search continues until most of the individuals occupy the same region of the search space. This region represents the local (and possibly global) optimum of the search space (Back & Schwefel, 1993; Davis, 1991; Liepins & Hillard, 1989; Michalewicz, 1992; Nissen, 1993; Whitley, 1993).

Evolutionary Algorithms as Optimization Methods

Evolutionary methods are regarded as weak optimization methods. Unlike strong optimization methods, weak optimization methods are not problem specific. Hence, for very specific problems, evolutionary algorithms are often outperformed by specialized (strong) optimization methods (Beasley et al., 1993a, b; Whitley, 1993; Mitchell, 1996). As Whitley (1993) states:

> [I]f there exists a good specialized optimization method for the specific problem, then genetic algorithm may not be the best optimization tool for application....

Yet, evolutionary algorithms and their derivatives have proved to be very efficient in solving a wide range of problems for which current optimization techniques are powerless. This is because:

- In contrast to other optimization methods, evolutionary algorithms work from a population of points on the problem space, not from a single point. Thus, they can search more complex problem spaces;
- Evolutionary methods use an efficient approach to the search of the problem space called an *implicit parallelism*. Thus, the evolutionary search is more efficient than the traditional one;
- Evolutionary methods use a pay-off (objective) function directly, not its derivative. Thus, they do not assume certain properties of an objective function (i.e., continuity and differentiability), often making the modeling more accurate;
- Evolutionary methods use probabilistic rather than deterministic transition rules. Thus, they are more flexible and realistic;
- Evolutionary methods work with the data about the problem itself, not with its parametric representation (Goldberg, 1989; Nissen, 1993). Their results are more accurate, since they make inferences about the data on the problem itself, not about the model of the problem.

Evolutionary Algorithms—Closer Encounters

In this section we shall discuss evolutionary algorithms (EA) in more detail. We shall begin with the presentation of the biological background of evolutionary algorithms. This presentation will extend beyond the brief discussion of natural evolution provided in the introductory part of the chapter and it will include some information on the cell reproduction and the composition of the DNA. We shall follow with an outline of an abstract model of a genetic evolutionary algorithm and with a brief discussion of different forms of this algorithm. We shall then progress to a discussion of components of an evolutionary algorithm and concepts of search space and fitness landscape. Towards the end of the chapter we shall present the schemata theorem—one of the theoretical models of evolutionary algorithms. We shall also discuss various measures of performance of evolutionary models. The chapter will conclude with the presentation of a simple evolutionary algorithm written in Java. Armed with this knowledge of evolutionary computations we shall progress to the second part of the book, which will discuss evolutionary models of spatial phenomena.

Biological Background

In our work with evolutionary algorithms (also called evolutionary computational models) we observed that the analogy between evolutionary computational models and mechanisms of biology and nature is a rather tricky business. Pushing this analogy too far clouds the real nature of evolutionary algorithms. On the other hand, not providing any, or providing too little information on the biological background of evolutionary algorithms, makes them look like any other computer creation—and the grand vision of the evolutionary paradigm standing behind the idea of these algorithms is lost. The following treatment of biological bases of evolutionary algorithms is, we feel, a necessary compromise between these two extremes. We hope that, after this brief introduction to the mechanics of natural evolution, the reader will be able to see how close (or how far) evolutionary computer algorithms are to the processes that they took their name from.

We can look at natural evolution on three levels: population, cell, and molecular. When viewed on a population level, natural evolution is concerned with the explanation of mechanisms of adaptation of individuals and populations to their environment. Viewed on the cell level, it is concerned with the processes of reproduction at the cellular level. Viewed on a molecular level, it is concerned with the molecular bases of reproduction and heredity.

As we shall see, the analogy between evolutionary computation and natural evolution is strongest on the population and cellular levels. When we go deeper into the molecular bases of natural evolution, this analogy becomes more and more distant.

Two observations originated the theory of natural evolution:

- The observation that similar species of animals with different features coexist in time, in different natural conditions (white rabbits in the North, brown rabbits in the South);

- The observation from the study of fossils that there is an apparent continuum of animal forms, from the long-gone species to the contemporary ones (for example, the transition of species from reptiles to birds or from reptiles to mammals and humans[19]).

The theory of natural evolution unifies these two observations and convincingly explains the mechanism underlying the observed continuity of patterns in species in space and time. The theory states that under the specific environmental conditions, organisms with particular combination of features suited for these conditions become dominant in the population. The features responsible for the domination of organisms become gradually, but very slowly, more pronounced. The accumulation of these small changes in the organism's features may produce, given sufficient time, a completely new species.[20]

The basis of the process of natural evolution is selection, a process in which some organisms or, more specifically, units of selection, survive (and mate and reproduce) and others do not. Selection, the main engine behind evolution, does not have a single expression or definition. Several types of selection have been observed:

- Normalizing selection with an effect of preserving the phenotype already present (this type of a selection does not further the evolution of an organism);
- Directional selection with an effect of favoring some particular combination of features in an organism;
- Diversifying or disruptive selection with an effect of producing several new phenotypes adapted to different environments.

Normalizing and directional selections tend to limit or maintain the variation of phenotypes in a population. Diversifying or disruptive selection tends to increase this variation.

In our discussion so far, we have used three important terms: *population, phenotype*, and *units of selection*. In order to avoid any confusion, it will be useful at this point to clarify the meaning of these terms. In the context of natural evolution, *population* refers to a group of organisms that interbreed or may interbreed. *Phenotype* is described as an outward expression of the combination of features of an organism. *Unit of selection* is an entity that, we think, the process of selection acts on. The unit of selection (a topic discussed in the introductory sections of this chapter) may be a gene, a group of genes, or a group of organisms. Recent evidence suggests that natural selection may operate on many levels of biological organization, from a single allele to groups of genes, to an individual organism and, finally, to groups of individuals (Hopson & Wessells, 1990).

The key mechanism of natural evolution is the transfer of features from parent to offspring organisms. This phenomenon is known as heredity. The mechanism of heredity relies on the fact that the features of both parents are stored in their genetic material (chromosomes) and get transferred from parent to offspring in the reproduction process.

There are two types of a cell reproduction processes: mitosis and meiosis. Mitosis takes place in single cell organisms, simple animals, and in somatic cells of complex organisms. Meiosis takes place in cells of complex organisms that have diploid (double) chromosomes.

In mitosis, the genetic material of the cell (we call this cell a parent cell) is duplicated and the cell divides itself into two daughter cells. Each daughter cell gets one exact copy of the parent's genetic material. Hence, the two daughter cells are genetically identical to themselves and to the parent cell. We say that these cells are clones. Mitosis perfectly preserves the genetic material of the parent and as such cannot produce any changes in the features of the offspring (barring some random mutations in the cell genetic material). The recombination of parental features and parental genetic material can only be carried out in meiosis.

Meiosis is the cell reproduction process in organisms with diploid cells, cells having two sets of chromosomes. Organisms with diploid cells are those organisms that reproduce via sexual reproduction. In a sexual reproduction process, a new organism is created by the fusion of reproductive cells (egg and sperm) from each parent. These reproductive cells are called gametes. From the point of view of genetics, each parent cell (each gamete) is effectively a half-parent, that is, it has only half of the total chromosome set of the complete cell. Gametes are what the meiosis process is generating and are what evolution is all about. Let us see why and how.

Diploid cells have pairs of chromosomes. Each pair of chromosomes contains one chromosome from one parent and another chromosome from another parent. These are called homologous pairs and each individual chromosome is called a homologue. If diploid cells were to reproduce through the process of mitosis, the new (daughter) cells would have twice the number of the chromosomes of the parent cell. The next reproduction cycle would again increase the number of chromosomes in daughter cells by two, and so on. But when gametes are fused, the new cell has exactly the same number of chromosome pairs as do parent cells, with one chromosome from each parent in each chromosome pair (a homologue).

Meiosis is more complicated than mitosis. It takes place in reproductive organs of an organism and it occurs in two stages, each stage to some extent resembling mitosis. Before meiosis begins as a part of a normal cell life cycle (so-called stage S) DNA is replicated and each chromosome has now two identical threads of DNA—daughter strands (chromatids)—joined together.

In the first stage of meiosis replicated pairs of sister chromosomes (chromatids) join together into tetrads. The tetrads are pairs of homologous chromosomes; each chromosome in a tetrad has two DNA threads (chromatids). When pairs of chromosomes are joined in tetrads, the segments of chromosomes between non-sister chromosomes are exchanged. This exchange of segments of chromosomes is called cross-over. The first stage of meiosis is completed when the tetrads separate and two new cells are created, each cell having the full set of randomly assigned duplicated chromosomes.

In the second stage, each of two new cells divides again. The sister chromosomes separate and each new cell gets one strand of the chromosome from every chromosome pair. New cells have a half of the full suite of chromosomes of the parent cells. These haploid cells (cells with single chromosomes) are gametes that are fused in the reproduction process. The fused cell has a full paired set of chromosomes of an original organism, but due to the exchange of the genetic material during meiosis and the reshuffling of chromosomes between the parent cells, the chromosome set of the fused cell is different from the chromosome set of the parent cells.

Exchange and reshuffling of the genetic material between parent cells happens several times during meiosis. First, the chromosomes cross over in both parent cells. Then, tetrads are randomly assigned to two cells in the first phase of meiosis. Finally, chromosomes are randomly assigned to gametes. And obviously, a particular sperm–egg combination is also a result of a random process, increasing the reshuffling of chromosomes (and genes).

In addition to the cross-over and recombination of chromosomes during meiosis there is another source of variation of the genetic material in the offspring. It is *mutation*—a random change of the chemical structure of a gene or the physical structure of a chromosome. The term mutation describes several distinct processes, two of which—*point mutation* and *chromosomal mutation*—are most important. A point mutation alters the properties of a single gene and creates a new allele. A chromosomal mutation rearranges blocks of genes on the chromosomes. These changes may include inversion, deletion, translocation, and duplication. The description of these processes is beyond the scope of this book and the interested reader may want to check existing references for more information on this topic (Hopson & Wessells, 1990).

Let us now discuss the molecular processes underlying reproduction and transfer of features between organisms. Chromosomes, carriers of the hereditary information, are composed of DNA (deoxyribonucleic acid), an extremely long molecule consisting of two strands of nucleotides wound around each other in a helix. It has been proven that the particular features of an organism are related to the specific segments of chromosomes. Those segments are called genes. Each gene representing the particular feature may have a different expression—called an *allele*. For example, a gene responsible for eye color may have alleles for brown, green, or blue eyes. Quite often, the particular feature is related not to one gene but to a group of genes, located on the same or different chromosomes. The complete genetic material of an organism is called a genotype. The part of a genotype that is expressed on the level of the organism is called a phenotype.

The mechanism of transfer of the hereditary information may be reduced to the process of synthesis[21] of amino-acids—molecules that form building blocks of proteins and eventually us. The DNA has encoded information on which amino-acids to synthesize and how to do this. To understand the synthesis of amino-acids we need to first look, even briefly, at the composition of DNA. The DNA molecule is composed of a chain of sequences of four nucleotides—adenine, guanine, cytosine, and thymine—that are coded as A, G, C, and T, respectively. Two strands of DNA are fused with hydrogen bonds between pairs of nucleotides A-T and G-C. The sequence of three nucleotides on a DNA strand forms one codon. One codon is used to synthesize one type of amino-acid. There are 4^3 (64) combinations of nucloid acids (codons) on the DNA. As there are 20 basic amino-acids, more than one type of codon is used to code for one amino-acid. Codons are arranged on a DNA string in a linear, non-overlapping fashion.[22] During the amino-acid synthesis the DNA sequence of nucleotides is used to assemble amino-acids molecules and proteins, and eventually organisms. We should note here that molecular processes are only distantly mirrored in computer

Table 1-6 Comparison of terminology of evolutionary algorithms and natural evolution (Goldberg, 1989a; Holland, 1993)

Natural Level	Interpretation	Algorithmic Level
Chromosomes	Strands of DNA	Sets of genes
Genes	A section of a chromosome responsible for a particular feature	String
Genotype	Totality of genetic material in an organism	Structure
Alleles	Values admissible to genes	Value encoded in the string
Mutation	Random changes to the chemical structure of a gene of to a physical composition of a chromosome	Random changes to a data structure representing a chromosome
Locus	Position of a gene in a chromosome	Gene index in multigene models
Phenotype	Detectable outward manifestation of a specific genotype	Point in problem space, or value of the objective function

evolutionary models. They are "the stuff" the DNA that computation is mimicking (Adleman, 1998).

This brief treatment of biological basis of natural evolution may not satisfy some readers. It can also be criticized for leaving out many key concepts in the vast topic of evolution. However, we hope that we were able to capture all these aspects of the evolutionary process that are necessary to derive the analogy with evolutionary computation models. We feel that more detailed coverage of this topic would cloud the explanation with unnecessary information. The interested reader is referred to many excellent books on this topic (Hopson & Wessells, 1990; Wallace, 1997).

To bridge the information on the biological mechanisms of evolution with our knowledge of computational evolutionary models, table 1-6 presents the comparison of analogous terms from these two domains.

One final issue requires further consideration. If our whole organism can be recreated from DNA, is DNA us? Instead of answering this question directly, let us give you an analogy that may be helpful in thinking about this dilemma. If I have a CD with Verdi's opera *Aida*, is the CD actually Verdi's opera? The answer is "of course not!" A CD is a medium on which the information about *Aida* is recorded. Is then DNA a CD on which something is recorded? If you think that this CD metaphor is inappropriate you may want to reconsider it. One of the basic principles of artificial intelligence (at least for now) is that knowledge can be represented by symbolic structures. Further, it is assumed that these symbolic structures can be represented on any medium. Thus, knowledge is independent of, and should not be confused with, the medium that encodes it (Cawsey, 1998; Russell & Norvig, 1995). The idea of information having its own existence is not as science-fiction-like as it may seem. The interested reader may consult the books by Barrow & Tipler (1988) and Stonier (1990), which discuss these issues in more detail.

At this point, we leave philosophical and biological issues and move on to the topic of computation.

Formal Representation of Evolutionary Algorithms

We shall progress now to the presentation of formal notation of basic concepts in evolutionary algorithms. Why do we need formal notation of evolutionary algorithms? The answer is quite simple. We need an unambiguous and compact method for representation of concepts and ideas. Evolutionary algorithms have acquired a unique formal representation. We shall restrict our presentation of formal notation of evolutionary algorithms to coverage of the most fundamental concepts.

Elements of evolutionary algorithms are formally represented as follows:[23] $f: \Re^n \rightarrow \Re$ is an objective function, $\Phi: I \rightarrow \Re$ is a fitness function, I is the space of individuals, $\vec{a} \in I$ is an individual. $\mu \geq 1$ denotes the size of the parent population, $\lambda \geq 1$ denotes the size of the offspring population. A population at a generation t is denoted as $P(t) = \{\vec{a}_1(t), \ldots, \vec{a}_\mu(t)\}$, where $\vec{a}_i(t) \in I$ are individuals in the population. Operators are defined as $O_{i\Theta_r}: I^\mu \rightarrow I^{i\mu}$ transforming a population I^μ into a population $I^{i\mu}$. Operators can be unitary or binary. Unitary operators act on one population and are defined as $U_{\Theta_u}: I^\mu \rightarrow I^\mu$ with parameters Θ_u. Binary operators act on two populations and are defined as $B_{\Theta_B} : I^\mu \rightarrow I^\mu$ with parameters Θ_u. A selection operator is defined as $s_{\Theta_s}: P(t) \rightarrow P(t+1)$. $\iota: I^\mu \rightarrow \{true, false\}$ is a termination operator defining the termination criterion. The termination criterion sets the condition that terminates the algorithm. Using this notation, a generic evolutionary algorithm may be described as follows:

$t \leftarrow 0$

initialize $P(0) = \{\vec{a}_1(0), \ldots, \vec{a}_\mu(0)\} \in I^\mu$

evaluate $\{\Phi(\vec{a}_1(o)), \ldots, \Phi(\vec{a}_\mu(o))\}$

select $s_{\Theta_s}: (P(0)) \rightarrow P(t)$

while $(\iota(P(t)) \neq true)$ do

  apply operator O_1

  ...

  apply operator O_n

  evaluate $\{\Phi(\vec{a}^{\,i}_1(t)), \ldots, \Phi(\vec{a}^{\,\iota}_\mu(t))\}$

  select $s_{\Theta_s}: (P(t)) \rightarrow P(t+1)$

  $t \leftarrow t+1$

endwhile

stop

Evolutionary Operators

Operators of initialization, selection, fitness, mutations, and cross-over are included in most implementations of EAs. These operators form the backbone of every EA. Without them, EAs would not be what they are.

Initialization Operator

Evolution starts with a population of individuals. We call it *initial population*. The individuals in the initial population are generated by an *initialization operator*. Each individual in the initial population represents an instance of the solution to a problem modeled by the evolutionary algorithm. Collectively, these individuals may be seen as points in the problem space. The individuals should cover the problem space as uniformly and completely as possible.[24]

The initialization operator defines the domain of an evolutionary model, data structure that represents individuals, and coding of individuals. The domain of the evolutionary model is the domain over which the modeled problem is defined. For example, if the evolutionary model represents a function of a single variable X, the domain of the evolutionary model (and of the initialization operator) is a range of values of $X[X_{min}, X_{max}]$. For the evolutionary model of a function defined by two variables X and Y, the domain of the model (and again of the initialization operator) are ranges of X and Y. If the evolutionary model represents a set of symbols, a computer program, a network, or a graph, the collection of these elements constitutes the domains for the particular problem and the domains of the particular evolutionary model. Consequently, initialization operators for the particular evolutionary model have to be given, as their domain, the particular set of symbols, a set of allowable computer instructions, or a set of network or graph elements.

Data structures that represent individuals are strings holding binary-coded values, strings holding naturally coded values, networks, trees, or tables. In complex evolutionary models, such as spatial evolutionary models, these data structures may represent *spatial objects*.

Coding of individuals refers to the way in which the problem is expressed in an evolutionary model. We shall discuss this concept more extensively in the subsequent parts of the book.

The core of an initialization operator is a random number generator. A random number generator should be robust, that is, it should be able to generate long sequences of pseudo-random numbers. It should also be able to accept a seed that changes a sequence of numbers every time the generator is restarted. Only a robust random number generator is able to generate the initial set of individuals that will uniformly cover the domain of the evolutionary model. A good and reliable random number generator is not an easy thing to find. If you believe that you found one, you should be cautioned by Knuth (1981):

> Anyone who considers arithmetical methods of producing random digits is, of course, in a state of sin.

Fitness Operator

Fitness assigned to each individual in a population by a *fitness operator* is a measure of how well a particular individual satisfies the objectives of the evolutionary model. Fitness of an individual is also the only link of the evolutionary computer model with the "real phenomena" that it purports to represent. In our initial example of an evolutionary algorithm ($f(x) = x^2$), fitness of an individual was calculated as a squared distance to the known optimum of the function $f(x) = x^2$. In real models, however, we usually cannot calculate fitness in this manner as we never know the value of the best solution of the model. The only thing we do know is how to differentiate a poor solution from a better one. In these cases fitness reflects our understanding of the structure of the problem, as well as the knowledge of what constitutes an acceptable solution. Needless to say, if the fitness function is designed incorrectly, the evolutionary algorithm will generate poor solutions and the whole evolutionary model will fall apart.

A fitness score drives the selection of individuals; the higher the fitness of an individual the more chances (sometimes many times more) that individual has to be selected for reproduction and to pass some of their genes to the next generation.

A good evolutionary algorithm tends to maintain high variability of individual fitness scores during the initial stages of evolution. However, quite often, the selection process eliminates lowest scoring individuals and fitness of individuals in the population converges to the same value after only a few generations. This fast convergence of a population to a single fitness value (so-called *loss of variability in a population*) often prevents an evolutionary algorithm from finding the best possible solution to the model. To facilitate more gradual convergence of fitness scores in a population and, consequently, to improve the chances of generating a better evolutionary model, several transformation methods of fitness scores have been developed. Most of these transformation techniques use some measure of the overall population fitness to compare a fitness score of a given individual with fitness scores of other individuals in the population.[25] Among the most commonly used measures of overall population fitness are *minimum population fitness, average population fitness*, and *standard deviation of population fitness*.

The techniques for the transformation of fitness scores include: *ranking, linear scaling, power scaling*, and *sigma scaling* (Davis, 1991). In a ranking technique, fitness scores of all individuals in a population are ordered from highest to lowest and each individual is assigned a number that is the rank order of its fitness score.

In a linear scaling technique, each individual in the population is assigned a fitness score based on the transformation function $f_s = af_r + b$, where f_s is a transformed (scaled) fitness score, f_r is a raw fitness score, and a and b are scaling parameters. Parameters a and b are adjusted as needed in the model.

In a power scaling technique, each individual is assigned a fitness score based on the transformation function $f_s = f_r^{pow}$, where f_s is a transformed (scaled) fitness score and f_r is a raw (not scaled) fitness score. Parameter *pow* is adjusted as needed in the model (in most cases the parameters are adjusted by experimentation). Depending on the choice of a *pow* parameter, a power scaling technique may increase or decrease the difference between lowest and highest scores.

In a sigma scaling technique, raw fitness scores are transformed using an average and a standard deviation of a population fitness score. The following sigma scaling function was proposed by Forrest (1985):

$$f_s = f_r - (f_{av} - c\sigma)$$

where f_s is a scaled fitness score, f_r is a raw fitness core, f_{av} is an average fitness score of a population, c is a multiplication factor (usually set between 1 and 3), and s is a standard deviation of a population fitness score. A different version of this formula is presented by Tanese (1989):

$$f_s = \begin{cases} 1.0 + \dfrac{f_r - f_{av}}{2\sigma} & \text{if } \sigma \neq 0 \\ 1.0 \quad \text{if } \sigma = 0 \end{cases}$$

In linear, power, and sigma scaling techniques, f_s, f_{av}, and f_r may or may not be indexed to the generation index. Indexing adjusts fitness scaling as the average and standard deviation of a population fitness change. Other variants of sigma scaling are discussed by Michalewicz (1992).

Selection Operator

A *selection operator* selects individuals for mating and cross-over. Selection is a sampling process: we are sampling a population to select individuals that are thought to be the best for the next generation.

The simplest selection method selects individuals according to their fitness in the population. This method is called the *roulette-wheel selection* or stochastic sampling with replacement. In this method, each individual is assigned a probability score that is obtained by dividing its individual fitness score by a total fitness score of the population. A random number is pooled and an individual with the fitness range that includes the pooled random number is selected for mating. For example, if the fitness f of an individual i is equal to 5.55 and the total fitness of all the individuals in a population is equal to 15.0, the fitness of f_i relative to the total population fitness f_{ir} is 5.55/15.0 = 0.37. We order relative fitness as $f_1, f_2, \ldots f_{n-1}, f_n$. The ith individual is selected if the pooled random number r from the range [0, ..., 1] is $f_{i-1} < r < f_i$. After each selection, an individual is returned to the population. In this selection method, individuals with very high fitness scores have very high probabilities of being selected several times during the selection process and consequently dominating the population. To prevent that from happening, various fitness scaling and transformation techniques are used (as described in the previous section).

The *Q-tournament selection* is a scheme in which q different individuals are selected from the population and the individual with the highest fitness score is retained. All individuals (including the selected one) are returned to the population. If $q = 2$, that is, if only two individuals are selected, the tournament is called 2-tournament (Back, 1996). This selection scheme has another variant in which following the initial selection of two individuals, a random number is pooled. If this number is greater than the value of a preset parameter k, the fitter of the two individuals is selected; otherwise, the other individual is selected. In both cases, individuals are returned to the original population (Mitchell, 1996).

The *elitist selection* retains some number of best individuals at each generation. In other selection methods the best individuals are quite often lost. They are not selected and their chromosomes, frequently containing desirable solutions to the modeled problem, are lost. The elitist selection method prevents these desirable individuals from being lost during the evolution.

The *rank selection* is based on the probability of selection derived from the rank order of an individual in the population, with ranks assigned according to the individual's fitness. The rank "1" goes to the individual with the best fitness score, the rank "2" goes to the second best, and so forth. Michalewicz (1992) provides the following formula for deriving this probability:

$$P_{select}(rank) = q(1 - q)^{rank-1}$$

where q is a user-defined parameter. For example, if $q = 0.1$, and the population size is 30, the probability of selection of the worst individual $P_{select}(30)$ is 0.004. $P_{select}(1)$ is 0.1.

In the majority of selection methods, the parent population is replaced by the offspring population. In contrast, *steady-state selection* is a method in which only the portion of the parent population is replaced by the offspring population. The number of replaced individuals is called a generation gap. Steady-state selection is most often used in evolutionary models involving learning or decision rules.

Cross-over Operator

A *cross-over operator* recombines genetic material between chromosomes of individuals. It takes as its input chromosomes of the parent organisms and produces the new set of chromosomes for the offspring organism. The chromosomes of an offspring organism are different than the chromosomes of parent organisms. Thus the features of the offspring organism differ from the features of the parent organisms.

The design of the cross-over operator closely reflects the design of an evolutionary model. If an evolutionary model manipulates strings, the cross-over operator exchanges segments of strings between chromosomes of individuals. A string may be separated into segments at one point (single-point cross-over) or at many points (multiple-point cross-over). Cross-over points may be chosen randomly (random-point cross-over) or may be fixed (fixed-point cross-over). If an evolutionary model manipulates numbers in their natural representation (numbers represented as n-byte reals or integers) a cross-over operator is implemented as a weighted average of chromosomes from two parent individuals (arithmetic cross-over). If an evolutionary model manipulates more complex data structures (trees, networks, grids), the cross-over operator is implemented as an operator specific to a particular data structure (trees, networks, grids, etc.). If an evolutionary model represents spatial phenomena using spatial objects, the cross-over operator is implemented as a spatial operator on spatial objects.

For example, for binary-coded strings or strings representing symbols, cross-over is implemented as shown in figure 1-2. In this example, cross-over enacts the exchange of segments (parts) of two chromosomes of different individuals: the chromosome [acpkia] and the chromosome [xaqajz]. As seen in the figure, during the cross-over four genes "pkia," from a chromosome on the left [acpkia], are exchanged with four genes "qajz," from a chromosome on the right [xaqajz], and new chromosomes [acqajz] and [xapkia] are created. This is obviously a single-point cross-over as the string-chromosome is segmented into single points.

$chromosome_1 = 55.02$, $chromosome_2 = 32.99$
cross-over of $chromosome_1$ and $chromosome_2$ is:

$$(a_1*55.02 + a_2*32.99)/(a_1 + a_2)$$

with $a_1 = 1$, $a_2, = 1$

and

$$0.5(55.02 + 32.99)$$

New chromosome is

$$44.005$$

Figure 1-4 A cross-over operation on two chromosomes with natural representation of numbers.

This cross-over operation can be implemented only if there are no "semantic" restrictions on the "symbols" encoded in the chromosome.

For numbers represented in natural coding the weighted arithmetic cross-over is implemented as shown in figure 1-4. If weights a_1 and a_2 are equal to 1, the cross-over is a simple arithmetic average of parent chromosomes.

For some complex structures representing complex objects the cross-over operator has to take into account the configuration of objects, as shown in figure 1-5. The structure represents an individual (symbolized by the oval) that consists of four objects. The four component objects of each individual each have defined specific locations on space. As we can see, there is no obvious or unique way to define cross-over for these two individuals. Cross-over that works with structures such as the one in figure 1-5 has been implemented for the spatial evolutionary algorithm described in the second part of this book.

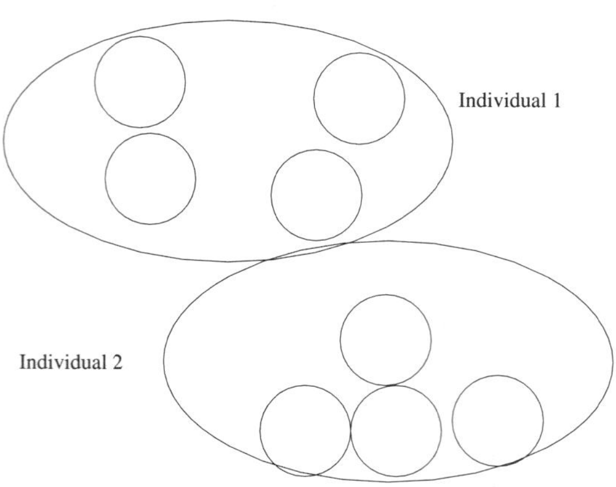

Figure 1-5 Cross-over between individuals representing complex data structures.

Of course, the cross-over operator must assure that individuals created by a cross-over operator belong to the problem domain (that they are legitimate from the point of view of the model): a task quite easy for the simple numerical representations, but quite involved for complex chromosomes representing limited sets of symbols (a set of rules), complex data structures, or complex semantic constructs such as computer programs.

Mutation Operator

A *mutation operator* randomly changes a composition of chromosomes. In contrast to the cross-over operator, which creates new chromosomes by recombining already existing structures, the mutation operator creates new chromosomes by adding new structures or new combinations of structures. These changes are random, that is, they are not guided by the structure of an evolutionary model.

The mutation operator must operate within the domain of the modeled problem. That is, the output of the mutation operator must belong to the domain of the problem and must be within the acceptable set of symbols, structures, or numbers for the particular representation of the problem. For example, if the chromosome is a binary string, mutation will flip a bit of the string at a randomly selected location on the chromosome, that is, it will change the bit from 1 to 0 or from 0 to 1 depending on the design of the mutation operator. If the chromosome represents a set of symbols, mutation will change the symbol at a randomly selected location on a chromosome to another randomly selected symbol from a set of permissible symbols.

The effect of mutation on the progress of an evolutionary algorithm may, or may not, be significant. In some implementations of evolutionary algorithms, mutation has been found to be a highly disruptive operator that did not contribute to the progress of an evolutionary algorithm at all (Krzanowski, 1997). In other implementations [such as in evolutionary programming (EP) or evolutionary strategies (ES)] it is the only operator causing changes in the chromosomes and consequently driving the progress of evolution. In canonical genetic algorithms (CGA), GP, and their derivatives, mutation is a part of an evolutionary process [along with the cross-over and (or) other operators] but its activation frequency is relatively low.

Other Operators

In addition to the selection, mutation, and cross-over operand, designers of evolutionary algorithms frequently introduce "problem-specific" operators. Among these problem-specific operators, of particular interest is the learning operator. The *learning operator* is defined as an operator that improves the fitness of organisms in a population between evolutionary cycles. The term "learning" as used in the evolutionary model[26] is defined after Ackley & Littman (1991) as "a process at the individual level whereby an organism becomes optimized for its environment."

The idea of learning within an evolutionary process has long been controversial (Hinton & Nowland, 1987). The debate on learning in natural evolution has focused on the question of whether learning could, potentially, influence the course of evolution. A central point in this debate has been an assumption that, in evolution, the transfer of features from one generation to another occurs through the exchange of genetic material. Learning, on the other hand, was assumed to act only upon the phenotype, without affecting the genetic make-up of an evolving unit. Consequently, according to this accepted school of thought, learning alone could not affect the genomes of future generations and therefore could not be regarded as an evolutionary mechanism.

However, some researchers (cf. Baldwin, 1986) suggested that, even if it does not directly affect the genome, learning may facilitate adaptation of an individual to a complex environment and in this way affect evolution. This view of learning as an adaptation to environment is known as the *Baldwin effect*.

In evolutionary algorithms, learning is usually implemented as some type of search algorithm, such as hill climbing, simulated annealing, taboo search, or any other search method fitting the particular modeled problem. Quite frequently, learning operators incorporated into evolutionary algorithms are very problem-specific, thus making the evolutionary algorithm a strong optimization method.

Studies demonstrated that addition of the learning operator substantially enhances the performance of evolutionary algorithms (Brady, 1985). In some cases evolutionary algorithms that incorporate learning can solve problems intractable otherwise (Ackley & Littman, 1991).[27]

Coding and Representation

The genetic make-up of organisms represents features of a modeled problem. The features in chromosomes can be represented as binary strings, numbers, symbols, or more complex structures. If features are represented in a chromosome as binary strings we call this representation method *binary coding*. If features are represented in a chromosome as strings, numbers, or more complex structures, we call this representation *natural coding*. For example, a chromosome representing a number 5, in natural coding, would contain a number "5." A chromosome using binary coding (we assume an 8-bit chromosome) would contain a string "00000101" for the same number.

The representation we are talking about here is the representation on the level of the high-level programming language used to write the code of the evolutionary algorithm. It is necessary to realize that these differences in representations are apparent, as all numbers, structures, symbols, and so forth, in a digital system are represented as 8-bit strings of "0"s and "1"s. This point will be re-emphasized at the end of this chapter.

Binary representation may be achieved by using the base 2 representation of the number, Gray coding (see an explanation later), or by mapping the range of numbers (or symbols) into the appropriate range of binary numbers. For example, using a base 2 representation, the number 2555 would be represented as "100111111011," a string of 12 bits, or the number 1,668,766 would be

Table 1-7 Comparison of binary and Gray-coded integers

Integer	Binary	Hamming Distance to Preceding	Gray	Hamming Distance to Preceding
2	0010		0011	
3	0011	1	0010	1
4	0100	2	0110	1
5	0101	1	0111	1
6	0110	2	0101	1
7	0111	1	0100	1
8	1000	3	1100	1

represented as "11001011101101001110." The range $[-2.0, 2.0]$ with 10,000,000 digits of accuracy we would represent with a string of 20 bits $(2^{20} > 1000000 > 2^{19})$. We map -2.0 into "00000000000000000000", and 2.0 into "11110100001001000000." The numbers in between -2.0 and 2.0 are mapped into the respective strings "00000000000000000001, 00000000000000000010, ..." One can easily see that if the evolutionary model that uses binary representation has to represent the very large numbers or a number of large numbers, the size of the corresponding chromosome may easily grow to the point of creating storage and manipulation problems.

Binary-coded representations are also affected by the so-called Hamming cliff problem (Goldberg and Dep, 1991; Whitley, 1993). The Hamming cliff can be best explained through the behavior of the mutation operator, although it affects all evolutionary operators on binary-coded genes. The mutation operator is usually designed to explore, by random perturbations, the vicinity of a chromosome in a problem space. If binary coding is used to represent an integer, the change of a single bit may result in a large change of value for a related integer—moving a chromosome out of the vicinity of initial location. For example, for a 3-bit chromosome [000] coding of an integer 0, a mutation affecting the first gene [100] would move the corresponding chromosome far away from the initial location—in fact, much farther than the mutations of either the third [001] or of the second [010] gene. In the first case the chromosome [100] would correspond to an integer 4—four units away from its initial location. In the second case the chromosome [001] would correspond to an integer 1—one unit away and the chromosome [010] would correspond to an integer 2—two units away.

The Hamming cliff problem can be overcome by using Gray coding. In Gray coding, adjacent integers always have a Hamming distance of 1, thus avoiding the problem of the Hamming cliff. Table 1-7 shows the Gray and binary codes for selected integers.

Thus, natural coding of chromosomes is a generally preferred coding method. Naturally coded chromosomes do not have to be decoded for calculation of an objective function, thus avoiding extra processing. Furthermore, natural coding avoids some problems that are inherent in binary coding, such as the Hamming cliff described above, which affects the representation of the large numbers and the large size of chromosomes.

Search Space and Fitness Landscape

The concepts of *search space* and *fitness landscape* are fundamental to understanding evolutionary processes and evolutionary algorithms. The following sections discuss these concepts and their role in the evolutionary process.

Search Space

The search space of an evolutionary algorithm refers to a set of all possible configurations of genetic material for a given evolutionary model (Mitchell, 1996). In other words, it is a set of all possible solutions to a problem represented by the evolutionary model. An evolutionary algorithm can be thought of as a *decision* algorithm that decides which elements of the search space to look at and in what order. Out of the set of all possible elements of the search space the best (i.e., optimal) solution is found. Thus, the size and the topography of the search space defines the complexity of the modeled problem faced by the algorithm.

Fitness Landscape

The fitness landscape of an evolutionary algorithm refers to the collection of all possible configurations of the genetic material with their fitness. The concept of the fitness landscape is important to the understanding of search mechanisms of an evolutionary algorithm as evolution (natural evolution as well as evolution simulated by an evolutionary algorithm) is the exploration of the fitness landscape. It has a specific meaning in spatial problems that will be explored later. The use of the term landscape only applies to a conceptual neighborhood.

Quite often, the complexity of the problem is described in terms of the topography of its fitness landscape (Mitchell, 1996). The fitness landscape is an n-dimensional function with n being equal to the number of different features represented by the chromosome. In a simplified case of only two dimensions, the fitness landscape may be compared to the model of a terrain elevation, with elevation representing fitness. The fitness landscape characterized by one single, highest peak corresponds to the optimal configuration of features in a chromosome. Alternatively, if there are several peaks of similar elevation, several suboptimal configurations of features are possible. The existence of suboptimal solutions creates several plateaus and valleys, making the fitness landscape more complex (reflecting the complexity of the problem).

Schemata Theorem

The schemata theorem explains the convergence process of canonical genetic algorithms (CGA) and their derivatives. The name of the theorem comes from a string (part of a chromosome) representing a set of subspaces of a search space of an evolutionary model. The schemata theorem states that an evolutionary algorithm may achieve convergence by exploring many short-length schemata, converging on the best ones and, thereafter, combining them into longer schemata (Holland, 1993; Davis, 1991; Whitley, 1993; Liepins & Hillard, 1989).

More precisely, a schema (or a template) is a set of bit strings (in binary-coded chromosomes) that can be described by a template composed of zeros, ones, and a "don't care" character—"*." The character "*" stands for either 1 or 0. For example, a string $H = 1 * * * 1$ is a schema of order two—it has two defined bits, and has a defining length 4 (bit 5 – bit 1 = 4). The string H describes a range of strings, such as 11111, 100*1, 10001. But it does not describe strings like 10000, 0000*, 01011. The string H is interpreted as representing a hyperplane of five dimensions in 0–1 coordinates with the first and the last (fifth) coordinates fixed. When an evolutionary algorithm evaluates a population of strings, it implicitly evaluates the average fitness of all strings for all the schemata present in the population—this is called implicit parallelism. Through selection and cross-over, an evolutionary algorithm increases the number of strings that have schemata with higher fitness, focusing the search on fewer and fewer hyperplanes. Now we can restate the schemata theorem as follows.

The number of specific schemata in the population $t + 1$ is proportional to the ratio of fitness of these schemata in the population t and the average fitness of populations, schemata defining length, and the probability of mutation where t is an index of a population ($t = 0$ for the initial population, $t = 1$ for the next population, etc.).

The schemata theorem can be succinctly expressed by the following formula:

$$E(m(H, t+1)) \geq \frac{u(H,t)}{f(t)} m(H,t)\left(1 - p_c \frac{d(H)}{l-1}\right)[(1 - p_m)^{o(H)}]$$

where t is an iteration of an evolutionary algorithm, H is a schema of length l, $m(H, t)$ is the number of schemata H at time t, $E(m(H, t + 1))$ is the expected number of schemata H at time $t + 1$, $f(t)$ is the average fitness of the population at time t, $u(H, t)$ is the average fitness of H at time t, $d(H)$ is the defining length of H, p_c is the probability of applying cross-over to string H, p_m is the probability of mutation, and $o(H)$ is the order of H (number of defined bits).

The formula can be interpreted as follows: the expected number of schemata H at time $t + 1$ is greater than or equal to the number of schemata H at time t times their average fitness $[u(H, t)/f(t)]$ $m(H, t)$ multiplied by probabilities that schemata H will survive cross-over $[1 - p_c\, d(H)/(l - 1)]$ and mutation $[(1 - p_m)^{o(H)}]$ (the independence of the cross-over and the mutation is assumed). The complete derivation of this formula can be found in Mitchell (1996).

It should be noted here that the validity of the schemata theorem and its underlying assumptions has been questioned and other theoretical models of genetic algorithms have been proposed (Mitchell, 1996).

Varieties of Evolutionary Algorithms

Classical (Canonical) Genetic Algorithms

The fundamental model of evolutionary algorithms is called the *canonical genetic algorithm (CGA)*. It was introduced by Holland (1993) and extensively described by several other authors (Beasley et al., 1993a; Brindle, 1980; Davis, 1991;

Goldberg, 1989; Liepins & Hillard, 1989; Whitley, 1993). CGAs have been designed as general models of evolutionary computations capable of solving numerical, classification, or decision problems.

The CGA represents the modeled problem as fixed-length, binary strings. Thus CGA operators do not work in the problem domain but on its binary representation and the effect of these operators on the model cannot be easily explained in terms of the problem space. The link between the problem space and its binary representations is done by special encoding and encoding schemes particular to the model. Closely associated with the binary representation used in CGAs is a concept of schemata and the schemata theorem, both explained in the preceding section.

The CGA uses three operators: the selection, mutation, and single-point cross-over. Subsequent developments brought more complex CGA models with multi-point cross-over and variable-length chromosomes.

While originally defined over binary, fixed-length strings, the CGA has been the source of inspiration for many other variants of evolutionary models, such as models manipulating symbols, rules, or objects. The CGA concepts are also at the core of the discussed in this book spatial evolutionary algorithms.

Evolutionary Programming

Originally proposed by Fogel (1962), evolutionary programming (EP) is a class of evolutionary algorithms designed to model an intelligent behavior. In EP, an individual is a finite state machine (FSM), represented as a state transition table (STT), with inputs, transition states, end state, and initial state. Fogel modeled individuals as FSMs as, according the computation paradigm of human cognition, an intelligent behavior could be reduced to computational processes with the FSM as their presentation.

The FSMs and their corresponding state transition tables represent systems that manipulate symbols according to some prescribed rules. Each input to the FSM leads to some "state." Figure 1-6 and table 1-8 present an example of a simple FSM. In this example, S_0 denotes the initial state. The input "1" moves the FSM to the state S_1. The input 0 moves the FSM to state S_2. To get to state S_3 from the state S_0 the FSM must receive the inputs "1,0" or "0,1." Other states in this FSM can be achieved in the similar way (Hopcroft & Ullman, 1979).

In the initial version of EP, the mutation operator was the only operator used. An individual could be mutated by changing the output symbol, the state transition, the number of states, or the initial state. The mutated individuals were tested for fitness. Individuals better than their parents were retained, while those worse than their parents were rejected.

Subsequent developments of EP techniques introduced different mutation schemes and different selection methods. As a consequence, EP models became quite similar to evolutionary strategies discussed in the next section. EP has been applied to pattern recognition problems, searching for optimum gaming strategies, and, recently, to continuous parameter optimization and neural network (NN) designs.

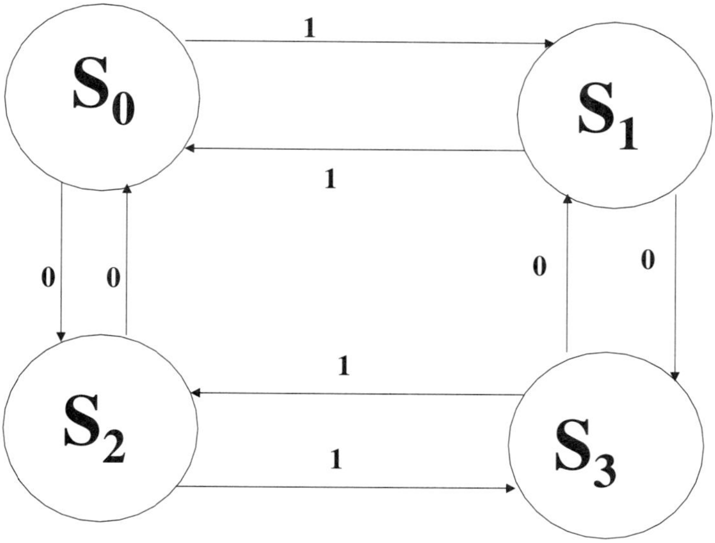

Figure 1-6 State transition diagram for a simple FSM.

Evolutionary Strategies

Evolutionary strategies (ESs) were originally developed during the 1960s (Schwefel, 1981; Back, 1996, Michalewicz, 1992). They are algorithms that have been specifically designed to solve parameter optimization problems. Early ESs used a floating-point number representation and the mutation as their only operator. These early ESs worked with only one individual in a parent and one individual in an offspring population. Mutation (and evolution) was realized by the formula:

$$X^{t+1} = X^t + N(0, \sigma)$$

where $N(0, \sigma)$ is a normal random variable with a standard deviation σ and X^t is an individual before mutation. The mutated individual was accepted as an

Table 1-8 State transition table for the state transition diagram in figure 1-6

	Inputs	
States	0	1
S_0	S_2	S_1
S_1	S_3	S_0
S_2	S_0	S_3
S_3	S_1	S_2

offspring only if its fitness was better than the fitness of the parent individual. Subsequent variations in this basic design added more parent population members, more offspring population members, different offspring population selection schemes, mutation operators with dynamically controlled parameters, and crossover operators. The different types of ES are represented by the following notation:

$$(\mu + \lambda)\text{-}ES$$

where μ symbolizes a number of individuals in the parent population and λ symbolizes a number of individuals in the offspring population. The plus sign indicates that the offspring population is selected from the temporary pool of individuals created by the offspring and the parent populations. Instead of the plus, the "-" sign is also used in denoting some types of ES designs and, when used, it indicates that the new population is selected only from among the offspring individuals. Using this notation, the original ES is represented as $(1 + 1)$-ES, which denotes a design with one parent, one offspring, and selection of the next population from a parent–offspring pair.

The ESs perform very well on numerical optimization problems—the type of problems for which they were originally designed.

Genetic Programming

Developed by Koza (1991), genetic programming (GP) is a class of evolutionary algorithms specifically designed to search for new computer programs. The representation of an individual in GP resembles the representations of the computer program in the computer. Three such representations are used in GP: a tree representing a parsing sequence for normal expressions, a linear array, and a graph. Using this form of representation, GP can model any phenomenon or construct that can be represented as a tree, or a graph. Thus GP can model simple phrases, a set of decision rules (what, in fact, is the computer program if not a set of decisions with a syntax!), a mathematical formula, or a dynamic n-dimensional data structures.

GP uses three operators: selection, cross-over, and mutation. Selection works in the same way as in other types of evolutionary algorithms. Cross-over works by exchanging fragments of trees (or whatever other structure GP is using for the given model) of different individuals. Mutation works by random modifications to the tree (or other structure) content. All operators must preserve the syntactic rules of the structure represented by GP.

The following example, taken from Banzhaf et al. (1998), may be helpful in clarifying the basic concepts behind GP. In this example, the objective of the presented GP model was to find out the function that best fits a set of 10 pairs of numbers generated using the function $f(x) = x^2/2$. The function was selected out of the set of permissible mathematical functions that included four operators: "+," "−," "∗," and "/." The domain of the function $f(x) = x^2/2$ was constrained to the interval $[-5.0, 5.0]$. The fitness score of an individual is an error between the values predicted by the function and those provided for the example

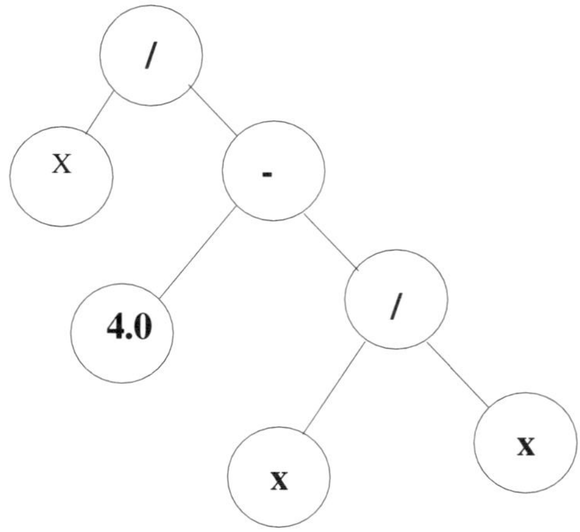

Figure 1-7 A simple tree representation of a mathematical formula, $[|x - 4.0|xx]$.

for a particular x. The initial population of individuals was composed of simple algebraic functions. One such function is given in figure 1-7.

In Polish notation this function can be expressed as $[/x - 4.0/xx]$; spaces have been added for clarity of presentation. The correct function for the modeled problem defined as $f(x) = x^2/2$ has been found by the GP algorithm in the third generation.

Genetic programming has been one of the most dynamic areas of research in the area of evolutionary algorithms. Applications of GP models have been reported in data mining, process and robot control, art, biotechnology, and in many aspects of computing sciences, ranging from computer security and data compression to derivation of cooperative strategies (Banzhaf et al., 1998). This list is not complete and is still growing.

Some see GP as a derivation or extension of the CGA. In our view such a classification is hardly justifiable. Although it is true that the CGA and GP share the same evolutionary concepts, they have been designed with different objectives in mind. They also use different sets of operators and representation schemes. However, one of the most important differences between the CGA and GP is the modeling concept behind these two methods. GP is designed to work on the actual representations of the modeled problem (e.g., parsing tree for language and computer instructions), whereas the CGA is designed to work on the secondary representations of the problem (e.g., fixed binary coded strings) of the original problem. These differences are significant enough to justify the claim that the GA and GP are two separate classes of evolutionary algorithms rather than derivatives of one another.

It should be easy to see that both GAs and GP on a certain level of abstraction are virtually the same: both use data structures (GA—strings, GP—trees or graphs) to represent the problem and both have selection, recombination, and mutation operators to manipulate these data structures (GA has operators designed to work on binary strings, GP has operators to work on trees and graphs). The use of the particular data structure is a matter of choice, let's say of convenience. The operator schemes differ only in implementation details. For example, the recombination operator exchanges fragments of data structures—substrings in the case of GA and subtrees in the case of GP. The whole discussion from which basic evolutionary model a particular evolutionary model has been derived may be missing a point. In our view all evolutionary methods are different facets, views, aspects of the same concept by different people at different times.

Hybrid Genetic Algorithms/Hybrid Evolutionary Algorithms

Out of countless experiments with evolutionary computational models a class of evolutionary algorithms called hybrid genetic algorithms (HGAs) or hybrid evolutionary algorithms (HEAs) has emerged. To understand the HGA or the HEA we have to look first at optimization methods in general. Broadly speaking, we have two classes of optimization methods: weak and strong. Weak optimization is a type of generic modeling methods that does not require a detailed problem specification and, therefore, can be applied to a variety of different modeling problems. In contrast, strong optimization methods require a detailed specification of the problem and consequently can be applied only to the problems that are well known and well defined (usually such problems are quite rare).

There is also a less obvious aspect of weak and strong optimization methods. Weak optimization methods do not need much problem-specific structural adjustment—with relatively small changes, the same methods can be applied to many problems. In contrast, strong optimization methods are problem-specific through and through. They reflect very closely the structure of the particular problem, or a class of problems, for which they have been designed. This specificity is the reason for both their success on the one hand, and a relatively limited use on the other.

Let us look at an example. If we want to calculate the amount of fuel needed for a spacecraft to get from the Earth's orbit to the Moon, we have a fairly well developed analytical formula to do it. We may say we have a strong optimization method to address this problem. In contrast, if we want to analyze the behavior of an ecosystem over time we have to rely on some approximate models or heuristics to provide us with some reasonable solutions, since no complete, closed analytical model for this problem exists. Here, we may say we do not have a strong optimization method to model an ecosystem, thus we must rely on weak optimization methods.

Depending on our understanding of the nature and structure of the problem in hand we will use different optimization methods. If we have to model a problem for which a fairly detailed description is possible, we would use one of the strong optimization methods that perform best on such tasks. But if we have to model a problem for which the detailed specification either does not exist or is extremely difficult we have to rely on the solutions provided by weak optimization methods.

But what about combining these two methods? Let a weak optimization method provide some initial solution to the problem when our understanding of the problem is limited and, then, let a strong method work on this initial, more specific solution and improve it. This is, indeed, a general idea behind the HEA.

HEAs are designed to extract the best from these two modeling methods. HEAs use weak evolutionary modeling approach to search for the initial solution when the problem is poorly defined, and they use strong modeling methods to improve on this suboptimal initial solution. Because strong optimization methods bring into the model the knowledge of the problem itself, the HEAs are frequently referred to as *knowledge-based* evolutionary algorithms.

Davis (1991) gives three rules for the design of the HEA:

1. In designing the HEA use the encoding that is problem-specific.
2. Incorporate into the HEA any features of the problem, the problem-specific algorithm, or domain-based heuristics.
3. Adapt the evolutionary operators to the new type of coding.

Hybridization schemes vary as widely as the problems to which the HEAs are applied. Extensive treatments of the HEA can be found in Fleurent & Ferland (1994), Al-Attar (1994), Medsker (1995), Davis (1991), Michalewicz (1992), and Lobo & Goldberg (1996).

A distinct class of the HEA are fuzzy genetic algorithms (FGAs). FGAs are genetic algorithms that use techniques and principles of fuzzy logic. The application of fuzzy logic principles in the FGA may range from the definition of fuzzy rules manipulated by the FGA, use of fuzzy operators, or use of fuzzy techniques for coding and representation. The best review of the FGA applications can be found in Pedrycz (1997).

Evolutionary Programs and Evolutionary Algorithms

Over the years of experiments and research with various forms of GA, GP, ES, HEA, and similar algorithms, new algorithms have been developed that are difficult to classify into any of these categories (Michalewicz, 1992; Back, 1996). For the lack of better terminology, new names cropped up, such as "non-standard GA," "modified GA," and "messy GA." It became obvious that all of these algorithms are very similar despite their differences (see the note 14 on GP) and they should be known by the same name. Thus, the class of *evolutionary algorithms* (Back, 1996) or *evolutionary programs* (Michalewicz, 1992) has been proposed.

According to Back (1996), evolutionary algorithms are "a class of direct, probabilistic search and optimization algorithms gleaned from the model of organic evolution. The main representatives of this computational paradigm, Genetic Algorithms (GA), Evolutionary Strategies (ES), and Evolutionary Programming (EP), which were developed independently of each other, are ... instances of a generalized Evolutionary Algorithm."

Michalewicz states that "To avoid all issues connected with classification of evolutionary systems, we call them simply—Evolution programs (EP^M)."[28]

One crucial element of the definition of EP^M, added by Michalewicz (1992), concerns the representation of the modeled problem. Michalewicz states:

> Classical genetic algorithms, which operate on binary strings, require a modification of an original problem into appropriate (suitable for GA) form; this would include

mapping between potential solutions and binary representation, taking care of decoders or repair algorithms, etc. This is not usually an easy task. On the other hand, evolution programs would leave the problem unchanged, modifying a chromosome representation of a potential solution (using "natural" data structures), and applying appropriate "genetic" operators. In other words, to solve a nontrivial problem using an evolution program, we can either transform the problem into a form appropriate for the genetic algorithm or we can transform the genetic algorithm to suit the problem. Clearly, classical GAs take the former approach, EP^Ms the latter.

Michalewicz favors the use of more generalized concept of evolution as a source of inspiration and analogy for computerized evolutionary models, exemplified in the paradigm of complex adaptive systems (CAS) (Holland, 1995). As he points out, the essence of evolutionary modeling is not in the type of coding or operators used but in the general processes of selection, recombination, and mutation occurring under the external pressure and providing the feedback mechanism to the evolving system. In this book we use the concept of the evolutionary algorithm in a sense proposed and defined by Michalewicz.

Performance Measures of Evolutionary Algorithms

An issue closely related to the problem of design of an evolutionary model is the question of determining the difference between poor and good models. Are two evolutionary models that provide similar results equally good? Does a particular evolutionary operator appreciably improve the performance of the evolutionary model? In order to answer these and similar type questions we need measures of performance.

Measures or metrics of performance of evolutionary algorithms may be grouped into two categories: *simple* and *complex*. Simple metrics are obtained as a direct measure of evolutionary search. Complex metrics, obtained by combination of several simple metrics, are comprehensive measures of the whole search process over the duration of evolution.

The category of simple metrics includes a number of evolutionary cycles T, the fitness of the best performing individual $u*(t)$, the fitness of the worst performing individual $w(t)$, and an average fitness of a population at a cycle $t - u*(t)$ (where t is an evolutionary cycle index). The category of complex measures include an on-line performance O_{np}, an off-line performance O_{fp}, an absolute performance A_{bf} and their normalized versions $[M]_c$, or $[M]_i$. The $[M]_c$ metric is normalized to the number of evolutionary cycles. The $[M]_i$ metric is normalized to the initial population fitness. The square brackets [ ] in these metrics symbolize normalization. Although simple measures are reasonably self-explanatory, complex ones need to be defined in greater detail.

On-line performance at time T, O_{np}, is an average fitness of all individuals over the number of cycles:

$$O_{np}(s, T) = ave(u(t)) \; t = 0, 1, \ldots, T$$

where T is a number of evolutionary cycles, s is a search space, $u(t)$ is a performance at an evolutionary cycle t, and $ave()$ is an average operator. Off-line

Table 1-9 A summary of metrics of evolutionary algorithms

Symbol	Type	Description
T	Simple	Number of cycles
$w(t)$	Simple	Worst performance
$u*(t)$	Simple	Best performance
$ave(u(T))$	Simple	Average performance at T
O_{np}	Complex	On-line performance
O_{fp}	Complex	Off-line performance
A_{bf}	Complex	Absolute performance
$[O_{np}]_{c/I}$	Complex	On-line performance relative to T or $u(0)$[a]
$[O_{fp}]_{c/I}$	Complex	Off-line performance relative to T or $u(0)$
$[A_{bf}]_{c/I}$	Complex	Absolute performance relative to T or $u(0)$

[a] $u(0)$ is a performance at cycle 0.

performance, O_{fp}, is defined as an average of best performances of individuals over the number of cycles:

$$O_{fp}(s, T) = ave(u*(t)) \quad t = 0, 1, \ldots, T$$

where T is a number of cycles, s is a search space, and $u*(t)$ is the best performance at a cycle t. Absolute performance is a difference between average initial and average terminating performance:

$$A_{bf} = u(T) - ave(u(0))$$

On-line performance, normalized by the number of cycles is given by the formula $[O_{np}(s, T)]_c = O_{np}(s, T)/T$. On-line performance normalized by the average fitness at $t = 0$ is expressed as $[O_{np}(s, T)]_i = O_{np}(s, T)/ave(u(0))$. Similar definitions can be given for:

- Normalized off-line performance, $[O_{fp}(s, T)]_c = O_{fp}(s, T)/T$ and $[O_{fp}(s, T)]_i = O_{fp}(s, T)/ave(u0))$
- Normalized absolute performance, $[A_{bf}]_c = A_{bf}/T$ and $[A_{bf}]_i = A_{bf}/ave(u(0))$.

Confronted with so many performance measures to choose from, we are often faced with the decision which measure should be used and in what circumstances. The following recommendations may help in this selection process.

Simple performance measures are directly related to the objective function of an evolutionary model and can be interpreted in terms of the objectives of the modeled problem. With these measures is it relatively easy to say which is a good, better, or the best performing evolutionary model as our reference point is the real-world solution offered by the model.

Complex performance measures are related to the efficiency of evolutionary search and are used to characterize evolutionary modeling process over several cycles. Complex performance measures are most useful in the design of adaptive search strategies or dynamic control mechanisms.

Normalized measures are used to compare the performance of evolutionary algorithms on different tasks or data sets. Table 1-9 summarizes the metrics of evolutionary search with their major characteristics.[29]

Table 1-10 Parameter values for selected evolutionary operators reported in literature

Operator	Mitchell	Grefenstette	Schaffer
Population size[a]	50–100	30–100	20–30
Mutation rate[b]	0.001	0.005	0.005–0.01

[a]Population size denotes here the number of individuals in one cycle. This is equivalent to the size of a hyperpopulation in spatial evolutionary algorithms.
[b]Mutation rate denotes the probability of mutation. This is equivalent to the probability of big mutation in spatial evolutionary algorithms.

Parameters of Evolutionary Algorithms

Precise criteria on how to set parameters of evolutionary algorithms have never been defined. In practice, parameters are often set by a trial-and-error approach (Mitchell, 1996). However, even with this approach, we have to start somewhere, that is, we need some initial estimates of parameter settings. In absence of other methodologies, the initial parameter settings can be quickly estimated using the recommendations from published research. These recommendations usually offer good initial estimates for the starting values of parameters for any model, even though they have been derived for the specific class of algorithms, or the class of problems. Table 1-10 presents the parameter values for mutation and population size for evolutionary models as reported by Mitchell (1996), Grefenstette (1986), and Schaffer et al. (1993).

A Code

There is no learning without examples. As the saying goes—read so you learn, practice so you know. Thus, having taken the reader through the introductory discussion on evolutionary algorithms and evolutionary models we are presenting a program code that embodies most of the discussed concepts. We want to venture the statement that any other evolutionary algorithm simply builds on this example. The code, taken from the book by Watson (1997)[30] on applications of artificial intelligence, is written in Java. The reader less familiar with Java might want to consult one of many excellent books on this language available in any respectable bookstore. We particularly recommend Weiss (1998) and Flanagan (1996). The complete listing of the code is given in tables 1-11a and 1-11b.

The code is a simple, yet complete, evolutionary model that belongs to the class of genetic algorithms. It uses binary coding, fixed-length chromosomes, and five operators—initialization, selection, mating, cross-over, and mutation. The model is initialized by specifying the number of chromosomes and the number of bits (genes) per chromosome. Selection is a mix of the elitist (the two highest ranking individuals are always preserved) and steady-state selection with a variable generation gap. Mating is random. Cross-over is random with a random, single cross-over point, and it is preceded by reshuffling of chromosomes. Mutation flips the bit of a randomly selected chromosome in a randomly selected location. The

Table 1-11a Java code of a genetic algorithm

```
// Genetic Algorithm Java classes
//
// Copyright 1996, Mark Watson. All rights reserved.

package mwa.ai.genetic;

import java.util.*;

public class Genetic extends Object {

  protected int NumGenes; // number of genes per chromosome
  protected int NumChrom; // number of chromosomes
  protected BitSet Genes[];
  protected float Fitness[];

  public Genetic() {
    System.out.println("In dummy Genetic constructor");
    NumGenes = NumChrom = 0;
  }

  public void init(int g, int c) {
    System.out.println("In Genetic::init(...)");
    Genes = new BitSet[c];
    for (int i=0; i<c; i++) {
      Genes[i] = new BitSet(g);
      for (int j=0; j<g; j++) {
        if (Math.random() < 0.5) Genes[i].set(j);
      }
    }
    NumChrom = c;
    NumGenes = g;
    Fitness = new float[c];
    for (int f=0; f<c; f++)     Fitness[f] = -999;
    Sort();
  }
  public boolean GetGene(int chromosome, int gene) {
    return Genes[chromosome].get(gene);
  }
  public void SetGene(int chromosome, int gene, int value) {
    if (value == 0) Genes[chromosome].clear(gene);
    else            Genes[chromosome].set(gene);
  }
  public void SetGene(int chromosome, int gene, boolean value) {
    if (value) Genes[chromosome].set(gene);
    else       Genes[chromosome].clear(gene);
  }
  public void Sort() {
    BitSet btemp;
    for (int c=0; c<NumChrom; c++) {
      for (int d=(NumChrom - 2); d>=c; d--) {
        if (Fitness[d] < Fitness[d+1]) {
          btemp = Genes[d];
          float x = Fitness[d];
```

Continued

```
        Genes[d] = Genes[d+1];
        Fitness[d] = Fitness[d+1];
        Genes[d+1] = btemp;
        Fitness[d+1] = x;
      }
    }
  }
}
public void DoCrossovers() {
  for (int m=0; m<NumChrom/2; m++) {
    CopyGene(m + NumChrom/2, m);
  }

  // copy the 2 best genes best genes so that their
  // genetc material is replicated frequently:
  for (int i=0; i<NumGenes; i++) {
    SetGene(NumChrom - 1, i, GetGene(0, i));
    SetGene(NumChrom - 2, i, GetGene(0, i));
    SetGene(NumChrom - 3, i, GetGene(0, i));
    SetGene(NumChrom - 4, i, GetGene(1, i));
    SetGene(NumChrom - 5, i, GetGene(1, i));
  }

  int num = NumChrom / 4;
  for (int i=0; i<num; i++) {
    int c1 = 2+ (int)((NumChrom - 2) * Math.random() * 0.99);
    int c2 = 2 + (int)((NumChrom - 2) * Math.random() * 0.99);
    if (c1 != c2) {
      int locus = 2 + (int)((NumGenes - 3) * Math.random());
      for (int g=0; g<locus; g++) {
        boolean temp = GetGene(c1, g);
        SetGene(c1, g, GetGene(c2, g));
        SetGene(c2, g, temp);
      }
    }
  }
}
// 'to' and 'from' are 'sorted' indices:
private void CopyGene(int to, int from) {
  for (int i=0; i<NumGenes; i++)
    if (GetGene(from, i)) SetGene(to, i, 1);
    else                  SetGene(to, i, 0);
}
public void DoMutations() {
  int c = 2 + (int)((NumChrom - 2) * Math.random() * 0.95);
  int g = (int)(NumGenes * Math.random() * 0.95);
  if (GetGene(c, g)) SetGene(c, g, 0);
else                 SetGene(c, g, 1);
}
// Override the following function in sub-classes:
public void CalcFitness() {
}
}
```

Table 1-11b Java code of the genetic algorithms

```
// File: textGenetic.java
// This file contains a text-mode test program
// for class Genetic.


public class testGenetic {

  static public void main(String args[]) {
    MyGenetic G = new MyGenetic();
    G.init(30,15); // 30 genes/chrom. 15 chrom in pop.
    for (int i=0; i<51; i++) {
      G.CalcFitness();
      G.Sort();
      if ((i%5)==0) {
        System.out.println("Generation " + i);
        G.Print();
      }
      G.DoCrossovers();
      G.DoMutations();
    }
  }
}

class MyGenetic extends Genetic {
  MyGenetic() {
    System.out.println("Entered MyGenetic::MyGenetic()\n");
  }
  public void init(int g, int c) {
    super.init(g, c);
  }
  public void CalcFitness() {
    for (int i=0; i<NumChrom; i++) {
      float fitness = 0.0f;
      for (int j=0; j<NumGenes; j++) {
        if (GetGene(i, j)) fitness += 1.0f;
      }
      Fitness[i] = fitness;
    }
  }
  public void Print() {
    for (int i=0; i<2; i++) {
      System.out.print("Fitness for chromosome ");
      System.out.print(i);
      System.out.print(" is ");
      System.out.println(Fitness[i]);
    }
  }
}
```

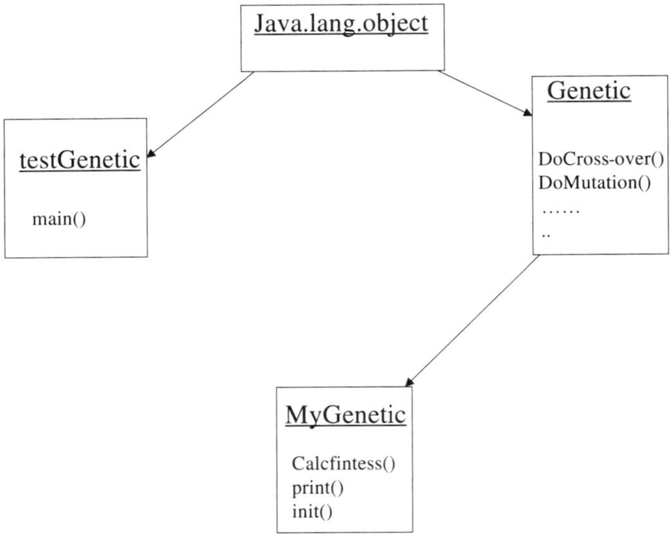

Figure 1-8 Object model of the genetic algorithm discussed in this section.

model also includes a less common operation in evolutionary models—reshuffling of chromosomes. The objective of the algorithm is to produce the highest fitness chromosome (chromosome composed entirely of ones) from the randomly generated chromosomes of zeros and ones. The chromosome fitness is the number of ones in the chromosome. The model runs for the predefined number of generations. No other termination criterion is given.

Walk-Through of the Code

The code consists of three classes: *genetic, mygenetic*, and *testgenetic*. The class *genetic* is a class that does most of the job. It includes two operator methods, cross-over (*docross-over*( )) and mutation (*domutation*( )), and three methods that manipulate genes, setgene( ), getgene( ), and copygene( ). The class genetic is extended by the class *mygenetic*. The class mygenetic includes print( ) and initialization, init( ), methods. The class *testgenetic* contains the main( ) method allowing the code to run as an independent application. The class hierarchy diagrams with the class names and methods are presented in figure 1-8.

Initialization

The initialization method creates an array of size c (number of chromosomes) of bitstrings of size g (number of genes) with each gene equal to 1 bit. Bits in the chromosomes have randomly allocated ones or zeros. The random number generator used in the algorithm is a method from the Java Math class. The method Math.random( ) generates pseudorandom numbers in the range [0.0, . . . , 1.0] (Chan & Lee, 1996). The code for the initialization method is given in table 1-12.

Table 1-12 Java code for the initialization method

```
public void init(int g, int c) {
  Genes = new BitSet [c];
  for (int i=0; i<c; i++) {
    Genes[i] = new BitSet(g);
    for (int j=0; j<g; j++) {
      if (Math.random() < 0.5) Genes[i].set(j);
    }
  }
```

The code uses the class BitSet. The BitSet class of java.util. BitSet defines an arbitrarily large set of bits and a set of methods to manipulate a bit set (Flanagan, 1996). In the code a class BitSet is used to create, maintain, and manipulate an array of bit sets; each bitset in the array is one chromosome.

An example of the initial population is given in table 1-13.

Fitness

Fitness of a chromosome is equal to the number of genes (bits) that have a value of "1." There is no scaling of fitness. The code of the fitness method and a sample population of chromosomes with the fitness scores are given in table 1-14. In table 1-15 the same population is sorted by fitness.

Table 1-13 Initial population

```
[000110110011001110011100100001]
[001000111100010100101110111100]
[001010000011100111010111010000]
. . .
[001010100111011111111010100100]
[100110010001101000101001100010]
[011011010001101101111111000000]
[101001110001101100000000111110]
[101110101110011000110100011110]
```

Table 1-14 Fitness method

```
public void CalcFitness() {
  for (int i=0; i<NumChrom; i++) {
    float fitness = 0.0f;
    for (int j=0; j<NumGenes; j++) {
      if (GetGene(i, j)) fitness += 1.0f;
    }
    Fitness[i] = fitness;
```

Table 1-15 Population sorted by fitness

```
Fitnes:14.0;[000110110011001110011100100001]
Fitnes:15.0;[001000111100010100101110111100]
Fitnes:13.0;[001010000011100111010111010000]
Fitnes:12.0;[100100000000011011011001011010]
Fitnes:15.0;[110110100001110010101010011010]
Fitnes:18.0;[111101100111010110100101110100]
Fitnes:9.0;[000001010000100010001000001111]
```

Reproduction

Reproduction starts with reshuffling of chromosomes: the first chromosome (the best) is copied to the chromosomes NumChroms −1, NumChroms −2, NumChroms −3, the second chromosome (second best) is copied to NumChroms −4, NumChroms −5. NumChroms is a variable holding the number of chromosomes. Then, two random numbers ($c1, c2$) are pooled and scaled to be within the range of [0 ... NumChroms −2]. The numbers $c1$ and $c2$ become indices of the chromosomes that are the candidates for cross-over. The third random number is pooled and scaled to be within the range of [0 ... NumGenes −3]. NumGenes is a variable holding the number of genes in a chromosome. The number $c3$ becomes a locus; a point of cross-over between chromosomes $c1$ and $c2$. The cross-over is implemented as an exchange of segments of two chromosomes. The chromosomes after the cross-over are replaced back into the population. The cross-over operation is done NumChroms/4 times. The code performing these operations is given in table 1-16.

This reproduction method is certainly not a typical one. However, it considerably speeds up the convergence of chromosomes, effectively eliminating chromosomes with low fitness scores already in the second generation.

Mutation

Mutation is always performed once in each generation. This is not typical for evolutionary models as in most evolutionary models (with an exception of ES and EP) mutation is activated only few times during the evolution. Mutation begins with pooling of two random numbers c and g. The number c is scaled to the number of chromosomes −2. The number g is scaled to the number of genes in a chromosome. Then the gene g on a chromosome c is flipped. If the gene g is 1, its value is changed to 0, if the g gene is 0 its value is changed to 1. The code for the mutation method is given in table 1-17.

Algorithm Workings

The typical sequence of events during the evolutionary run is presented in a program trace in table 1-18. The trace shows the sequence of activation of operators in generations 1, 2, 28, 27, and 50. It also shows which chromosomes have been crossed over at what locus and which chromosomes and genes have been

Table 1-16 Cross-over method of the genetic algorithm

```
public void DoCrossovers() {
  for (int m=0; m<NumChrom/2; m++) {
    CopyGene(m + NumChrom/2, m);
  }
  // copy the 2 best genes best genes so that their
  // genetc material is replicated frequently:
  for (int i=0; i<NumGenes; i++) {
    SetGene(NumChrom - 1, i, GetGene(0, i));
    SetGene(NumChrom - 2, i, GetGene(0, i));
    SetGene(NumChrom - 3, i, GetGene(0, i));
    SetGene(NumChrom - 4, i, GetGene(1, i));
    SetGene(NumChrom - 5, i, GetGene(1, i));
  }
  int num = NumChrom / 4;
  for (int i=0; i<num; i++) {
    int c1 = 2+ (int)((NumChrom - 2) * Math.random() * 0.99);
    int c2 = 2 + (int)((NumChrom - 2) * Math.random() * 0.99);
    if (c1 != c2) {
      int locus = 2 + (int)((NumGenes - 3) * Math.random());
      for (int g=0; g<locus; g++) {
        boolean temp = GetGene(c1, g);
        SetGene(c1, g, GetGene(c2, g));
        SetGene(c2, g, temp);
          }
      }
    }
  }
```

mutated. The trace allows us to look inside the evolutionary process and to see what the algorithm is actually doing.

The question of what happened to the chromosomes during the evolution is answered by the traces of populations at the generations 1, 5, 24, 40, and 50. The trace of each population contains a complete listing of the chromosomes in this population.

The population at the generation 1 is an initial population. The generations 5, 24, 40 represent transitory populations. The last population at the 50th generation presents the final solution of the algorithm.

The following symbols are used in the presentation of traces: "ch:0" is a chromosome number; "fitness;14.0" denotes the fitness core for a particular chromosome, [000110110011001110011100100001] is the chromosome. The index of genes runs from left, "0," to right, "29." The index of chromosomes runs from 0 to 14.

Table 1-17 Mutation method

```
public void DoMutations() {
  int c = 2 + (int)((NumChrom - 2) * Math.random() * 0.95);
  int g = (int)(NumGenes * Math.random() * 0.95);
  if (GetGene(c, g)) SetGene(c, g, 0);
  else         SetGene(c, g, 1);
```

Table 1-18 A typical sequence of operators activation in the run of the genetic algorithm

```
initial pop with genes and chromosomes: 30, 15 is created

Generation 1
  Cross-over between [4--9] at locus - 22
  Cross-over between [5--14] at locus - 6
  Cross-over between [12--7] at locus - 10
  mutated chromosome;14 mutated gene 21
Generation 2
  Cross-over between [11--8] at locus - 27
  Cross-over between [5--2] at locus - 14
  Cross-over between [8--9] at locus - 27
  mutated chromosome;12 mutated gene 6
Generation 27
  Cross-over between [10--8] at locus - 11
  mutated chromosome;6 mutated gene 19
Generation 28
  Cross-over between [8--7] at locus - 2
  Cross-over between [14--2] at locus - 3
  Cross-over between [8--12] at locus - 20
  mutated chromosome;5 mutated gene 11
Generation 50
  Cross-over between [10--9] at locus - 2
  Cross-over between [2--10] at locus - 18
  mutated chromosome;4 mutated gene 17
```

Table 1-19 presents the initial population. The same population sorted according to fitness is given in table 1-20.

Cross-over of two chromosomes (4 and 9) in the first population is given in table 1-21. Cross-over occurs at the locus (gene) 22. Before cross-over the chromosomes are reshuffled: a chromosome 0 has been mapped to 7, a chromosome 1 to 8, a chromosome 2 to 9, 3 to 10, and so forth. Then, the best chromosome is copied to the chromosomes 14, 13, and 12. The second best chromosome is copied to the chromosomes 11 and 10. Because of the reshuffling of chromosomes (see how reproduction is done in the model), a chromosome 9 is a chromosome 2 from the beginning of the generation. The population after the crossover is presented in table 1-22. As we can see, after the cross-over, low-scoring chromosomes are eliminated and the overall fitness of the population is increased.

Tables 1-23 through 1-26 present the populations at the generations 5, 24, 40, and 50 (final population) respectively. From the first to the fifth generation the population fitness increases quite fast. Subsequent generations improve fitness only slightly. This fast convergence of fitness scores is forced by reshuffling of chromosomes before each cross-over. The reshuffling preserves the best chromosomes and eliminates low-scoring ones.

As a word of caution we should mention here that in real-life models one is trying to avoid too quick convergence of fitness scores and slowing down of the evolution process. All of the previously discussed fitness scaling schemes were intended to do precisely that.

Table 1-19 Initial population of individuals

	Initial population is created with 15 chromosomes, each chromosome with 30 genes
ch;0Fitnes:14.0; [000110110011001110011100100001]	Population is not sorted on fitness
ch;1Fitnes:15.0; [001000111100010100101110111100]	
ch;2Fitnes:13.0; [001010000011100111010111010000]	
ch;3Fitnes:12.0; [100100000000011011011001011010]	The left column should be interpreted as follows: chromosome 3; fitness 12
ch;4Fitnes:15.0; [110110100001110010101010011010]	
ch;5Fitnes:18.0; [111101100111010110100101110100]	
ch;6Fitnes:9.0; [000001010000100010001000001111]	
ch;7Fitnes:13.0; [000000110010101110110011001010]	
ch;8Fitnes:20.0; [111111111110111110000101001010]	
ch;9Fitnes:13.0; [011101110011011001010000010000]	
ch;10Fitnes:17.0; [001010100111011111111010100100]	
ch;11Fitnes:12.0; [100110010001101000101001100010]	
ch;12Fitnes:16.0; [011011010001101101111111000000]	
ch;13Fitnes:14.0; [101001110001101100000000111110]	
ch;14Fitnes:17.0; [101110101110011000110100011110]	

Table 1-20 Initial population sorted by fitness

```
ch;0Fitnes:20.0;[111111111110111110000101001010]
ch;1Fitnes:18.0;[111101100111010110100101110100]
ch;2Fitnes:17.0;[001010100111011111111010100100]
ch;3Fitnes:17.0;[101110101110011000110100011110]
ch;4Fitnes:16.0;[011011010001101101111111000000]
ch;5Fitnes:15.0;[001000111100010100101110111100]
ch;6Fitnes:15.0; [110110100001110010101010011010]
ch;7Fitnes:14.0;[000110110011001110011100100001]
ch;8Fitnes:14.0;[101001110001101100000000111110]
ch;9Fitnes:13.0;[001010000011100111010111010000]
ch;10Fitnes:13.0;[000000110010101110110011001010]
ch;11Fitnes:13.0;[011101110011011001010000010000]
ch;12Fitnes:12.0;[100100000000011011011001011010]
ch;13Fitnes:12.0;[100110010001101000101001100010]
ch;14Fitnes:9.0;[000001010000100010001000001111]
```

Table 1-21 A cross-over of two chromosomes—"4" and "9"

before cross-over

ch;4Fitnes:16.0;[01101101000110110111111 *1000000*]

ch;9Fitnes:13.0;[**00101010011101111111101** 0100100]

after cross-over

ch;4;[**00101010011101111111101** *1000000*]

ch;9;[01101101000110110111111 0100100]

Table 1-22 A population after cross-over and mutation

X between [4--9] at locus - 22	Cross-over between chromosomes 4 and 9 at locus 22
X between [5--14] at locus - 6	Cross-over between chromosomes 5 and 14 at locus 6
X between [12--7] at locus - 10	Cross-over between chromosomes 12 and 7 at locus 10
ch;0;[111111111110111110000101001010]	
ch;1;[111101100111010110100101110100]	
ch;2;[001010100111011111111010100100]	
ch;3;[101110101110011000110100011110]	
ch;4;[001010100111011111111011000000]	
ch;5;[111111111100010100101110111100]	
ch;6;[110110100001110010101010011010]	
ch;7;[111111111110111110000101001010]	
ch;8;[111101100111010110100101110100]	
ch;9;[011011010001101101111110100100]	
ch;10;[111101100111010110100101110100]	
ch;11;[111101100111010110100101110100]	
ch;12;[111111111110111110000101001010]	
ch;13;[111111111110111110000101001010]	
ch;14;[001000111110111110000101001010]	
mutated chromosome;14 mutated gene 21	Mutation is activated; chromosome 14 and gene 21 are mutated
ch;0Fitnes:20.0;[111111111110111110000101001010]	
ch;1Fitnes:20.0;[111111111100010100101110111100]	
ch;2Fitnes:20.0;[111111111110111110000101001010]	
ch;3Fitnes:20.0;[111111111110111110000101001010]	
ch;4Fitnes:20.0;[111111111110111110000101001010]	
ch;5Fitnes:18.0;[111101100111010110100101110100]	
ch;6Fitnes:18.0;[111101100111010110100101110100]	
ch;7Fitnes:18.0;[111101100111010110100101110100]	
ch;8Fitnes:18.0;[111101100111010110100101110100]	
ch;9Fitnes:17.0;[001010100111011111111010100100]	
ch;10Fitnes:17.0;[101110101110011000110100011110]	
ch;11Fitnes:17.0;[011011010001101101111110100100]	
ch;12Fitnes:16.0;[001010100111011111111011000000]	
ch;13Fitnes:15.0;[110110100001110010101010011010]	
ch;14Fitnes:14.0;[001000111110111110000001001010]	

Table 1-23 A population at the generation 5

```
ch;0Fitnes:24.0;[111111111111110110110101111100]
ch;1Fitnes:24.0;[111111111111110110110101111100]
ch;2Fitnes:24.0;[111111111111110110110101111100]
ch;3Fitnes:24.0;[111111111111110110110101111100]
ch;4Fitnes:24.0;[111111111111110110110101111100]
ch;5Fitnes:22.0;[111111111110110110110101110100]
ch;6Fitnes:22.0;[111111111110110110110101110100]
ch;7Fitnes:22.0;[111111111110110110110101110100]
ch;8Fitnes:22.0;[111111111110110110110101110100]
ch;9Fitnes:22.0;[111111111110110110110101110100]
ch;10Fitnes:22.0;[111111111110110110110101110100]
ch;11Fitnes:22.0;[111111111110110110110101110100]
ch;12Fitnes:21.0;[110111111110110110110101110100]
ch;13Fitnes:21.0;[111111111110110110100101110100]
ch;14Fitnes:21.0;[111111111110110110100101110100]
```

Table 1-24 A population at the generation 24

```
ch;0Fitnes:27.0;[111111111111111111110111111100]
ch;1Fitnes:27.0;[111111111111111111110111111100]
ch;2Fitnes:27.0;[111111111111111111110111111100]
ch;3Fitnes:27.0;[111111111111111111110111111100]
ch;4Fitnes:27.0;[111111111111111111110111111100]
ch;5Fitnes:27.0;[111111111111111111110111111100]
ch;6Fitnes:27.0;[111111111111111111110111111100]
ch;7Fitnes:27.0;[111111111111111111110111111100]
ch;8Fitnes:27.0;[111111111111111111110111111100]
ch;9Fitnes:27.0;[111111111111111111110111111100]
ch;10Fitnes:27.0;[111111111111111111110111111100]
ch;11Fitnes:27.0;[111111111111111111110111111100]
ch;12Fitnes:27.0;[111111111111111111110111111100]
ch;13Fitnes:27.0;[111111111111111111110111111100]
ch;14Fitnes:26.0;[111111101111111111110111111100]
```

Table 1-25 A population at the generation 40

```
ch;0Fitnes:27.0;[111111111111111111110111111100]
ch;1Fitnes:27.0;[111111111111111111110111111100]
ch;2Fitnes:27.0;[111111111111111111110111111100]
ch;3Fitnes:27.0;[111111111111111111110111111100]
ch;4Fitnes:27.0;[111111111111111111110111111100]
ch;5Fitnes:27.0;[111111111111111111110111111100]
ch;6Fitnes:27.0;[111111111111111111110111111100]
ch;7Fitnes:27.0;[111111111111111111110111111100]
ch;8Fitnes:27.0;[111111111111111111110111111100]
ch;9Fitnes:27.0;[111111111111111111110111111100]
ch;10Fitnes:27.0;[111111111111111111110111111100]
ch;11Fitnes:27.0;[111111111111111111110111111100]
ch;12Fitnes:27.0;[111111111111111111110111111100]
ch;13Fitnes:27.0;[111111111111111111110111111100]
ch;14Fitnes:26.0;[111111111111110111110111111100]
```

Table 1-26 A population at the generation 50

```
ch;0Fitnes:28.0;[111111111111111111110111111110]
ch;1Fitnes:28.0;[111111111111111111110111111110]
ch;2Fitnes:28.0;[111111111111111111110111111110]
ch;3Fitnes:28.0;[111111111111111111110111111110]
ch;4Fitnes:28.0;[111111111111111110110111111110]
ch;5Fitnes:28.0;[111111111111111111110111111110]
ch;6Fitnes:28.0;[111111111111111111110111111110]
ch;7Fitnes:28.0;[111111111111111111110111111110]
ch;8Fitnes:28.0;[111111111111111111110111111110]
ch;9Fitnes:28.0;[111111111111111111110111111110]
ch;10Fitnes:28.0;[111111111111111111110111111110]
ch;11Fitnes:28.0;[111111111111111111110111111110]
ch;12Fitnes:28.0;[111111111111111111110111111110]
ch;13Fitnes:28.0;[111111111111111111110111111110]
ch;14Fitnes:27.0;[111111111111111111110111111110]
```

The presented example of a genetic algorithm is simple but it has all the key elements that make every evolutionary algorithm. The next part of the book will show how these principles can be applied to design a new model of evolutionary algorithm—a spatial evolutionary algorithm.

Notes

1. The term "spatial phenomena" refers to any process or system that occupies some space. It may be a polymer molecule, a crystal, a VLSI chip, a computer network, a city, a highway network, etc. The term "geographic" refers to the subset of spatial phenomena that are located in a geographic space. A more detailed explanation of what "space" is, is given in the second part of the book.

2. We are referring to the statement by J. Bronowski from his book *The Origins of Knowledge and Imagination*: "our knowledge of the outside world depends on our models of perception" (Bronowski, 1978). Another very interesting account of the growth of modern physical models of the universe and our view of reality through our models is given in an old, but still fascinating, book written by two giants of science, Albert Einstein and L. Infeld—*The Evolution of Physics* (Einstein & Infeld, 1938).

3. The struggle between religious and scientific views of the universe are discussed with typical Russellian vigor, cynicism, and lucidity in his book *Religion and Science* (Russel, 1961).

4. Probably the most penetrating account of the Copernican revolution, which is what is being discussed here, as well as its impact on society, is given by T.S. Kuhn in his work *The Copernican Revolution* (Kuhn, 1957).

5. An easily understood account of the development of models of the universe is given by I. Ekerland in *Mathematics and the Unexpected* (Ekerland, 1990). A masterful—though somewhat difficult to digest—account of the rise and fall of scientific paradigms, our views on reality and associated tensions is offered by Kuhn in his book *The Structure of Scientific Revolution* (Kuhn, 1996). Equally interesting (and equally difficult), but well worth reading is an account of how our views (models) of reality affect our thinking and our life, is given by Margolis in his book *Paradigms and Barriers* (Margolis, 1993). And what could be a better source of information about the struggle of a scientific prophet shackled by powerful forces within his society than the writer's own account of his struggle? For this, the reader is invited to an unparalleled masterpiece of dialectics by the master himself, Galileo Galilei and his letters to the Grand Duchess Christina (Finocchario, 1989).

6. A very prophetic view of the information flood is expressed by Openshaw in his book on new models in Geography, *Artificial Intelligence in Geography* (Openshaw & Openshaw, 1997). Openshaw is not alone in his opinions. But most of the authors perceive the "*information glut*" as a negative aspect of the current era of information technology. For those opinions see Shenk (1997), Rawlins (1996) and Postman (1993). For a historian's analysis of the information flood and its effects on our lives and societies see Roszak (1994).

7. Symbols, rules to manipulate the symbols, plus the rules of mapping a problem to these symbols constitute the model.

8. The full title of Darwin's work is more revealing: *The Origin of Species by Means of Natural Selection or the Preservation of Favored Races in the Struggle for Life* (Darwin, 1859).

9. Models based on mathematical expressions generally require the expression of the problem in the language of the model. If the problem is too complex, this representation is very difficult.

10. Tacit assumptions behind the model may include, but are not limited to, normality, linearity, and additivity of components, convexity of objective functions, continuity (*eo ipso* differentiability) of the problem space, and an existence of the optimal solution.

11. The genesis of the postulate that the reality can be adequately represented by a formal logic can be traced to both Aristotelian and Newtonian views of the world; each today reflected in the cognitive and computer sciences. For details of the postmodern critique of this, and a counter-position, see Raper (2000).

12. By "normal" we mean phenomena conforming to normal or Gaussian probability density distribution.

13. Modeling methods that belong to this new class of models are *simulated annealing, neural networks*, and *DNA computing*.

14. In fact, various forms of evolutionary modeling methods have been researched independently of complex adaptive systems (CAS) concepts (Back, 1996). However, over the years CAS principles have been recognized as being fundamental to all of these models.

15. We can substitute the term "evolutionary algorithm" for "GP" in this quote.

16. It is an obvious simplification of biology but accepted in evolutionary modeling.

17. We use the term "set" not in a strict mathematical sense. It is rather a collection of elements.

18. In this illustration, the term "individual" is used synonymously with the term "chromosome" as in the presented model each individual has only one chromosome.

19. More technical details about the past biological life on Earth can be found in Hopson & Wessells (1990). The book by Johnson (1993) about evolution makes interesting reading. While Johnson differs with Darwin, reading his contrarian views makes you go over the fundamental concepts of the evolution with more attention to details that you would do while reading less controversial works.

20. This is actually the principle behind the natural evolution. As Darwin himself states: "Natural selection can act only by the preservation and accumulation of infinitestimally small inherited modifications, each profitable to the preserved being . . . ."

21. A very complex process!

22. The DNA strand and a linear coding of codons resembles a Turing machine and a paper tape, using the analogy of Adleman (1998).

23. The abstract notation for evolutionary algorithms was proposed by Back & Schwefel (1993). It forms the basis for the expanded notation used for spatial evolutionary algorithms defined in this work.

24. There are reported successful applications in which the initial population included known suboptimal solutions to the modeled problem.

25. The relative performance measures are used in evaluations of performance of many systems or phenomena. For example, it is not important how high a student's SAT scores are on the absolute scale. What counts is what these scores are in comparison to, let's say, the national average.

26. A different definition of learning (supervised and unsupervised) is used in the context of classifiers.

27. For more details on evolutionary algorithms and learning see Mitchell (1996).

28. The superscript M is the author's addition to distinguish Michalewicz's EP from Fogel's EP.

29. For an overview of evolutionary metrics in nonspatial evolutionary models see Mitchell (1996); see Brindle (1980), Grefenstette (1986), or DeJong (1975) for on-line, off-line, and normalized measures.

30. The book by Watson (1997) contains the interesting example of an application of EA to rule learning in the computer strategy game—an example worth studying by itself.

References

Ackley, D. & M. Littman. 1991. Interactions between learning and evolution. In: C.G. Landgton, C. Taylor, J.D. Farmer, & S. Rasmussen (eds.), Artificial Life II, SFI Studies in the Sciences of Complexity: Vol. 10, 487–507. Engelwood, N.J.: Addison-Wesley.

Adleman, L.M. 1998. Computing with DNA. Scientific American, (8), 54–61.

Allander, J.T. 1995a. An indexed bibliography of genetic algorithms basics. Report series No. 94-1-Basics. Department of Information Technology and Industrial Management. University of Vaasa, Finland.

Allander, J.T. 1995b. An indexed bibliography of genetic algorithms in power engineering. Report series No. 94-1-Power. Department of Information Technology and Industrial Management. University of Vaasa, Finland.

Allander, J.T. 1995c. An indexed bibliography of genetic algorithms and fuzzy logic. Report series No. 94-1-Fuzzy. Department of Information Technology and Industrial Management. University of Vaasa, Finland.

Allander, J.T. 1995d. An indexed bibliography of genetic algorithms in control. Report series No. 94-1-Control. Department of Information Technology and Industrial Management. University of Vaasa, Finland.

Allander, J.T. 1995e. An indexed bibliography of genetic algorithms and artificial intelligence. Report series No. 94-1-AI. Department of Information Technology and Industrial Management. University of Vaasa, Finland.

Allander, J.T. 1995f. An indexed bibliography of genetic algorithms in CAD. Report series No. 94-1-CAD. Department of Information Technology and Industrial Management. University of Vaasa, Finland.

Allander, J.T. 1995g. An indexed bibliography of evolutionary strategies. Report series No. 94-1-ES Department of Information Technology and Industrial Management. University of Vaasa, Finland.

Allander, J.T. 1995h. An indexed bibliography of genetic algorithms in operations research: Years 1985–1994. Report series No. 94-1-OR. Department of Information Technology and Industrial Management. University of Vaasa, Finland.

Al-Attar, A. 1994. A hybrid GA-heuristic search strategy. AI Expert, 10, 34–37.

Back, T. 1996. Evolutionary Algorithms in Theory and Practice. New York: Oxford University Press.

Back, T. & H.-P. Schwefel. 1993. An overview of evolutionary algorithms for parameter optimization. Evolutionary Computation, 1 (1), 1–23.

Baldwin, J.M. 1986. A new factor in evolution. American Naturalist, 30, 441–451.

Banzhaf, W., P. Nordin, R.E. Keller, & F.D. Fracone. 1998. Genetic Programming—an Introduction. San Francisco: Morgan Kaufmann.

Barrow, J.D. & F.J. Tipler. 1988. The Anthropic Cosmological Principle. New York: Oxford University Press.

Beasley, D., D.R. Bull, & R.R. Martin. 1993a. An overview of genetic algorithms: Part 1, Fundamentals. University Computing, 15 (2), 58–69.

Beasley, D., D.R. Bull, & R.R. Martin. 1993b. An overview of genetic algorithms: Part 2, Research Topics. University Computing, 15 (4), 170–181.

Brady, R.M. 1985. Optimization strategies gleaned from biological evolution. Nature, 371 (31), 804–806.

Brindle, A. 1980. Genetic algorithms for function optimization. Unpublished doctoral dissertation, University of Alberta.

Bronowski, J. 1978. The origins of knowledge and imagination. New Haven: Yale University Press.

Cawsey, A. 1998. The Essence of Artificial Intelligence. New York: Prentice Hall.

Celko, J. 1993. Genetic algorithms and database indexing. Dr. Dobb's Journal, 4, 30–34.

Chan, P. & R. Lee 1996. The Java Class Libraries. An Annotated Reference. Reading, Mass.: Addison-Wesley.

Darwin. C. 1859. The Origin of Species by Means of Natural Selection or the Preservation of Favored Races in the Struggle for Life. Beckenham, Kent: Down.

Davis, L. (ed.) 1991. Handbook of Genetic Algorithms. New York: Van Nostrand Reinhold.

Davis, N. 1996. Europe. London: Oxford University Press.

Dawkins, R. 1976. The Selfish Gene. New York: Oxford University Press.

DeJong, K.A. 1975. Analysis of the behavior of a class of genetic adaptive systems. Unpublished doctoral dissertation, University of Michigan.

Delahaye, D., J.-M. Alliot, M. Schoenauer, & J.-L. Farges. 1994. Genetic algorithms for partitioning air space. In: The 10th Conference on Artificial Intelligence for Applications, 291–297. San Antonio, Texas. Los Alamitos: IEEE Computer Society Press.

Einstein, A. & L. Infeld. 1938. The Evolution of Physics. New York: Touchstone Press.

Ekerland, I. 1990. Mathematics and the Unexpected. Chicago: University of Chicago Press.

Finocchario, M.A. (ed.). 1989. The Galileo Affair. A Documentary History. Berkeley: University of California Press.

Flanagan, D. 1996. Java in a Nutshell. Sebastopol, Australia: O'Reilley & Associates.

Fleurent, C. & J.A. Ferland 1994. Algorithms genetic hybrids pour l'optimisation combinatoire. Montreal: Université de Montreal (in French).

Fogel, D.B. 1991. System Identification Through Simulated Evolution: A Machine Learning Approach to Modeling. Needham Heights, Mass.: Ginn Press.

Fogel, L.J. 1962. Autonomous automata. Industrial Research 4, 14–19.

Forrest, S. 1985. Scaling fitness in the genetic algorithm. In: Documentation for prisoner's dilemma and norms programs that use the genetic algorithms. Unpublished manuscript.

Gell-Mann, M. 1988. The Quark and the Jaguar: Adventures in the Simple and the Complex. New York: Freeman.

Goldberg, D.E. 1989. Genetic Algorithms in Search, Optimization, and Machine Learning. Reading, Mass.: Addison-Wesley.

Goldberg, D.E. & K. Dep. 1991. A comparative analysis of selection schemes used in genetic algorithms. In: G.J.E. Rawlings (ed.), Foundations of Genetic Algorithms, 69–79. San Mateo, Cal.: Morgan Kaufmann Publishers.

Grefenstette, J.J. 1986. Optimization of control parameters for genetic algorithms. IEEE Transactions on Systems, Man, and Cybernetics, Jan./Feb., 122–128.

Hinton, G.E. & S.J. Nowland. 1987. How learning can guide evolution. Complex Systems, 1, 495–502.

Hofstadter, D.R. 1979. Goedel, Escher, Bach: an Eternal Golden Braid. New York: Vintage Books.

Holland, J.H. 1993. Adaptation in Natural and Artificial Systems, 2nd Ed., Cambridge, Mass.: MIT Press.

Holland, J.H. 1995. Hidden Order. Reading, Mass.: Helix Books.

Holland, J.H. 1998. Emergence. From Chaos to Order. Reading, Mass.: Helix Books.

Hopcroft, J.E. & J.D. Ullman. 1979. Introduction to Automata Theory, Languages and Computation. Reading, Mass.: Addison-Wesley.

Hopson, J.L. & N.K. Wessells. 1990. Essentials of Biology. New York: McGraw-Hill.

Hosage, C.M. & M.G. Goodchild. 1986. Discrete space location-allocation solutions from genetic algorithms. Annals of Operations Research, 7, 35–46.

Isaaks, E.H. & M.R. Srivastava. 1990. An Introduction to Applied Geostatistics. New York: Oxford University Press.

Knuth, D.E. 1981. The Art of Computer Programming, 2nd Ed. Reading, Mass.: Addison-Wesley.

Koza, J.R. 1991. Genetic Programming. Cambridge, Mass.: MIT Press.

Krzanowski, R.M. 1997. Evaluation of spatial evolutionary algorithm for spatial modeling. Unpublished doctoral dissertation. University of London.

Kuhn, T.S. 1957. The Copernican Revolution. Cambridge, Mass.: Harvard Press.

Kuhn, T.S. 1996. The Structure of Scientific Revolution. Chicago: University of Chicago Press.

Liepins, G.E. & M.R. Hillard. 1989. Genetic algorithms: foundations and applications. Annals of Operations Research, 21, 31–58.

Lloyd, E.A. 1993. The Structure and Confirmation of Evolutionary Theory. Princeton, N.J.: Princeton University Press.

Lobo, F.G. & D.E. Goldberg 1996. Decision making in a hybrid genetic algorithm. IlliGAL Report No. 96009. University of Illinois.

Margolis, H. 1993. Paradigms and Barriers. Chicago: University of Chicago Press.

Maturana, H.R. & F.J. Valera. 1992. The Tree of Knowledge. New York: Random Century House.

Medsker, L.R. 1995. Hybrid Intelligent Systems. Boston, Mass.: Kluwer Academic.

Michalewicz, Z. 1992. Genetic Algorithms + Data Structures = Evolutionary Programs. New York: Springer-Verlag.

Michalewicz, Z. 1993. A hierarchy of evolution programs: an experimental study. Evolutionary Computation, 1 (1), 51–76.

Mitchell, M. 1996. An Introduction to Genetic Algorithms. Cambridge, Mass.: MIT Press.

Nissen, V. 1993. Evolutionary algorithms in management science. Papers on Economics and Evolution #9303. Goettingen: University of Goettingen.

Openshaw, S. & C. Openshaw. 1997. Artificial Intelligence in Geography. New York: John Wiley & Sons.

Pedrycz, W. (ed.) 1997. Fuzzy Evolutionary Computation. Boston, Mass.: Kluwer Academic Publishers.

Poincaré, H. 1952. Science and Hypothesis. New York: Dover.

Postman, N. 1993. Technopoly. New York: Vintage Press.

Raper, J.F. 2000. Multidimensional Geographic Information Science. London: Taylor & Francis.

Rawlins, G.J.E. 1996. Moths to the Flame. Cambridge, Mass.: MIT Press.

Roszak, T. 1994. The Cult of Information. Berkeley: University of California Press.

Russel, B. 1961. Religion and Science. New York: Oxford University Press.

Russell, S. & P. Norvig. 1995. Artificial Intelligence. A Modern Approach. Englewood Cliffs, N.J.: Prentice-Hall.

Schaffer, J.D., R.A. Caruana, & R. Das. 1993. A study of control parameters affecting online performance of genetic algorithms. Proceedings of the 9th Conference on AI for Applications, Orlando, Florida, 18–25.

Schwefel, H.-P. 1981. Numerical optimization of computer models. Chichester: John Wiley.

Shenk, D. 1997. Data Smog: Surviving the Information Glut. San Francisco: Harper.

Simon, H.A. 1996. The Science of Artificial Intelligence, 3rd Ed. Cambridge, Mass.: MIT Press.

Sober, E. 1993. The Nature of Selection. Chicago: University of Chicago Press.

Stonier, T. 1990. Information and the Internal Structure of the Universe. London: Springer-Verlag.

Swirski, P. 1997. A Stanislaw Lem Reader. Evanston: Northwestern University Press.

Tanese, R. 1989. Distributed Genetic Algorithms for Function Optimization. Unpublished doctoral dissertation, University of Michigan.

Wallace, R.A. 1997. Biology. New York: Benjamin Cummings.

Watson, M. 1997. Intelligent Java Applications. San Francisco: Morgan Kaufmann.

Weiss, W.A. 1998. Data structures and Problem Solving Using Java. New York: Addison-Wesley.

Whitley, D. 1993. A genetic algorithm tutorial. Technical Report CS-93-103. Ford Collins: Colorado State University.

PART II

SPATIAL EVOLUTIONARY MODELING

Algorithms and Models

2

Modeling Spatial Phenomena

In part II we describe some possible methods of modeling spatial phenomena with spatial evolutionary algorithms. We will explain what spatial evolutionary models and spatial evolutionary algorithms are and how they can be designed. We will also provide a general framework for spatial evolutionary modeling. We believe that this framework can be used to create evolutionary models (and algorithms) of spatial phenomena that will reach well beyond the model discussed in the book. Wherever possible we will give examples to illustrate the concepts, terms, and procedures we discuss. In fact, by the end of part II we will have built, using presented principles, a complete spatial evolutionary model—a spatial evolutionary model of a wireless communication system.

We shall begin our discussion with an explanation of the distinction between spatial evolutionary models and evolutionary models of spatial phenomena. As we shall see, the difference between these two terms, while subtle, is very important for the understanding of spatial modeling in general and evolutionary spatial modeling in particular.

"Spatial Evolutionary Models" Versus "Evolutionary Models of Spatial Phenomena"

The differences between the terms *spatial evolutionary models* and *evolutionary models of spatial phenomena* extend well beyond their lexical dissimilarities and touch upon very basic issues of evolutionary and spatial modeling. The term *spatial evolutionary model*, as used here, refers to an evolutionary model that constitutes a separate, distinct class of computer evolutionary models. In contrast, the term *evolutionary models of spatial phenomena* denotes applications of existing evolutionary methods (or mere extensions of established evolutionary methodologies) to problems defined in space. Our view of the science of spatial modelling is driven by the choice of which definition, along with its consequences, that we accept.

If we accept that spatial evolutionary models constitute a separate and distinct class of evolutionary models, then we will also have to accept the proposition that they possess unique rules governing their behavior, a unique genome design to represent a model-specific data structure, and a set of unique operators that cannot be readily applied to nonspatial problems.[1] Moreover, it will follow that these evolutionary models also possess problem-specific *language*, that is language specific to the domain of spatial evolutionary models.[2]

However, the claim that we have developed a new class of models may be met with controversy. Alternatively, when we propose that evolutionary models of spatial problems are merely extensions of "traditional" nonspatial evolutionary methods with no special design requirements, such a postulate is much less contentious and, as a result, also much less likely to arouse protest. Many in the field of spatial information systems, and in the field of geographic information systems (GISs) in particular, will find it difficult to resign themselves to the idea that spatial problems do not require any distinct modeling mechanism, or that spatial problems can be effectively modeled with the same methods as nonspatial problems. Such a view simply defies their everyday experience. It is for that group of individuals (and we include ourselves among them) that this book has been written.

The authors wish to defer any conclusive judgment as to where spatial evolutionary methods belong within the domain of evolutionary models. We do believe, and will attempt to illustrate it in this book, that spatial evolutionary models are distinct from other forms of evolutionary algorithms. We think that spatial evolutionary models require a modeling approach that is very specific to the domain of spatial problems. Moreover, we also think that problems modeled by spatial evolutionary methods are qualitatively different from problems for which nonspatial models are applied. Indeed, we believe that almost all aspects of spatial evolutionary models are sufficiently unique to warrant the claim that spatial evolutionary algorithms constitute a distinctly separate class of evolutionary models.[3] We shall reiterate this claim at the end of this chapter after the discussion of an example of the spatial evolutionary model.

Spatial Evolutionary Models: What Are They?

A collection of associated objects forms a system. When these objects are located in space, we call this system a *spatial system*. We define a spatial system as a combination of selective physical elements and patterns of associations that they form.[4] Some examples of spatial systems are:

- a city;
- a network of highways;

- weather patterns;
- a palette load;
- a complex molecular structure of a protein;
- a mineral crystal;
- a computer chip;
- a computer network with servers, routers, and network connections;
- a partition of space into "subspace" according to some criterion, such as the partition of airspace for purposes of air traffic control;
- a partition of a mineral deposit into blocks of different ore grade;
- a set of facilities serving some spatially distributed demand, such as a network of automatic teller machines (ATMs), banks, retail stores, or schools.

The following description of one such spatial system—a city—is taken from the book *The Timeless Way of Building* by Alexander (1979):

> On the geometric level, we see certain physical elements repeating endlessly, combined in an almost endless variety of combinations. A town is made of houses, gardens, streets .... Each urban region is defined by certain patterns of relationships among its elements. Evidently, a large part of the "structure" of a building or a town consists of patterns of relationships. At first sight, it seems as though these patterns of relationships are separate from the elements. When we look closer, we realize that these relationships are not extra, but necessary to the elements, indeed a part of them. When we look closer still, we realize that even this view is still not very accurate. For it is not merely true that the relationships are attached to the elements: the fact is that the elements themselves are patterns of relationships.

All spatial systems, whatever they represent, share the same four common, basic characteristics:

1. Spatial systems are composed of spatial phenomena;
2. Spatial phenomena have properties;
3. Spatial phenomena are interrelated;
4. Spatial systems have characteristics—metrics—that allow for differentiation between several similar systems.

For example, a network of roads (spatial system) is composed of roads (spatial objects). A road has properties such as name, number, and traffic capacity. The roads are intersecting and interconnecting, forming a pattern of associations. A network of roads can be characterized by their combined length, or combined traffic-carrying capacity. Table 2-1 provides several more examples of spatial systems, as viewed from a perspective of the four-point framework.

We write about spatial models so we are supposed to take it for granted what space is. However, the definition of space is actually quite complex. A casual glance at the index of Russell's book *On Human Knowledge* (Russell, 1948) shows that space has several meanings depending on who is talking. There may be an absolute space, physical space, common-sense space, metric, Euclidean, non-Euclidean, topological, objective, subjective, perceived, inferred, perceptual, psychological, private, and public space. And there is also a geographic space. Even in this book we have had other spaces also: problem space, solution space, model space. So what space are we talking about here? When we talk about spatial models we mean a physical, or an absolute space. As the concept of the

Table 2-1 Four characteristics of spatial systems and examples of their interpretations

Characteristic	Examples of Features
Spatial systems are composed of spatial objects	A network of highways is composed of highways Cities are composed of streets, buildings, parks, etc.
Spatial objects have designated properties	Highways have length, size, type, number Buildings are addresses, names, locations
Spatial objects are interrelated	Highways are connected, and intersect each other Buildings are close, far near, face each other
Spatial system has designated characteristics that allow for differentiation between several similar systems	Highway network has highway density, traffic capacity, etc. Cities have number of buildings, percent of free space, etc.

absolute or physical space may not be obvious we will give several definitions of it, hoping that the reader will find at least one of them clear.

Whitehead (1920) defined the absolute space as follows: "[s]pace ... is the outcome of certain relations between objects commonly said to be in space; and whenever there are the objects, so related, there is the space." And a bit later he said more succinctly: "[S]pace is an abstraction from the relations between material objects." Newton, as quoted by Russell (1948), said that "The space consists of the collection of points, each devoid of structure, and each one of the ultimate constituents of the physical world." Newton's definition is very close to the definition of space (space is a collection of points) we use in mathematics. Einstein in his foreword to Jammer's book on space (Jammer, 1993) gave two (complementary) definitions of space: "a) space is a positional quality of the world of material objects; b) space is a container of all material objects." There was also a common-sense definition of space, which goes like this: space is what is left after you take out all of the objects (in it). The definitions of space are important to us, spatial scientists, as they define the essence of our trade and in fact they constitute the conceptual foundations of our representation and our models (including computer models) of space. On specifics of space in the context of geography you may consult Worboys (1995) or Couclelis (1992). The work by Tuan (1977), *The Space and Place*, also makes interesting reading about space and the experience of the space.

Let us define some of the terms we use in our discussion of spatial systems. We will start with the generic concept of *spatial phenomenon*, a basic object of spatial modeling (and, of evolutionary spatial modeling as we will see later). A spatial phenomenon is a real-world phenomenon defined in space and time. There is not room in this book to deal with the philosophical assumptions and commitments that underlie this starting point. See Raper (2000) for a comprehensive discussion of the philosophical and methodological background to spatial and temporal representation.

The smallest spatial phenomenon capable of discrete representation can be called a *spatial entity*. Spatial entities can be grouped into classes. Classes of spatial entities have some common characteristics. Classes of entities that are defined for the purpose of modeling are called entity types. An instance of the entity type is called an entity instance. Both phenomena and entities have attributes and relationships but in this book we will use the term phenomenon to refer to a conceptual domain and the term entity to refer to the represented domain. The attributes describe qualitative or quantitative properties of phenomena or entities. The relationships describe the association between entities or phenomena.

The following example should help to clarify these concepts. A spatial phenomenon can be conceived relative to a philosophical and methodological position such as realism; an example could be a bridge. Let us consider, then, the George Washington (GW) Bridge, which spans the Hudson River in New York. This phenomenon (the GW bridge) belongs to the class of phenomena "bridge concepts" that are conceived in the same way. We can discuss and reason about such "bridge concepts" in a conceptual domain without geometric representation. Now, we want to create a spatial model involving geometric representation; our model will be a map of crisp entities, which are New York bridges, where the conceptual domain and the represented domain are mapped to each other in a defined way. The GW Bridge becomes an entity instance the moment we decide to represent it as a part of the model.

Another example of spatial phenomena is a river, which belongs to the class of phenomena "river concepts." Thus, the Hudson River, if we decide to use it in our spatial model in the represented domain, becomes an entity instance. Now we can place the Hudson River and the GW Bridge entities in several relations with each other. There is a relation "under" between the river and the bridge, as well as relations "over," "across," "spanning," and many others.

Represented spatial entities, when made computable in digital systems, can be called spatial objects. The basic spatial objects are points (zero dimensional), lines (one dimensional), polygons (two dimensional), and volumes (three dimensional). Spatial objects may be simple or complex. A complex spatial object is a combination of simple spatial objects. An entity may be represented by one or more simple or complex spatial objects.

Let us go back to our example. In a digital system, our two computable entities, the Hudson River and the GW Bridge, can be represented by one or more spatial objects. The Hudson River can be represented either as a line, a polygon with the designated left and right sides and an area in between, or as a set of two lines designating two river banks. The GW Bridge can be described in a similar way.

The process of transformation, abstraction, or mapping of physical phenomena into entities and their implementation as spatial objects is summarized in the diagram below:

Real-world phenomena → Entities → Special objects

An example should clarify any ambiguities about the nature of this process. Let us look at the spatial model (the USGS topographic map) of a rural landscape from New Jersey. The map is shown in figure 2-1. On the map, several entities can be recognized by reference to the symbology used in the map: roads, rivers, lakes, mountains, and buildings. These entities can be used in our

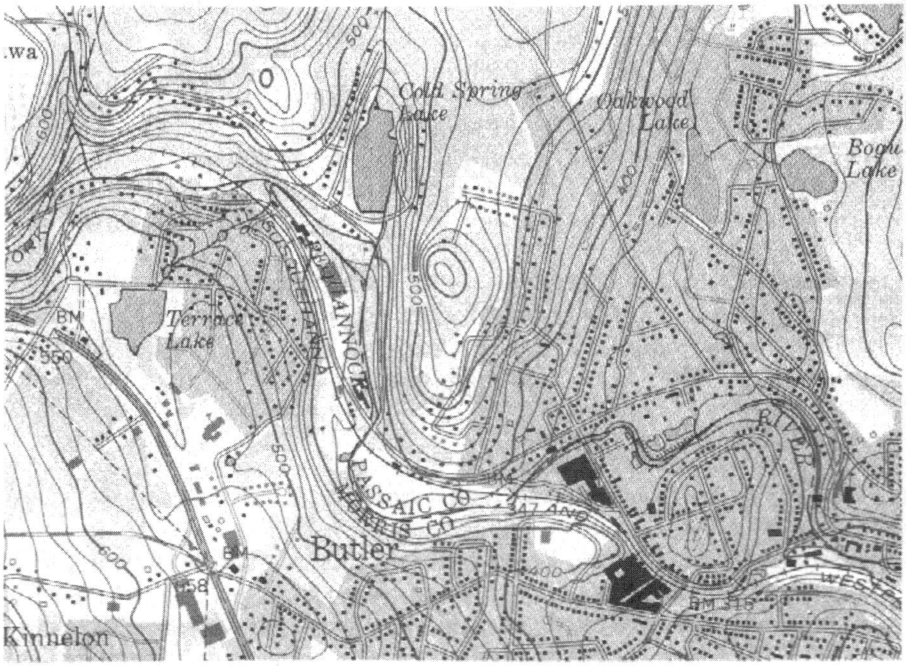

Figure 2-1 A USGS map of the area in New Jersey.

model, being represented by entity types such as classes of roads, buildings, and so forth. As the entities on the map have been abstracted from conceptualized phenomena for the specific purpose of the creation a model of the landscape—they are entity instances. Each building, road, river, and so forth, is a specific instance of the entity type selected for our model. Our entity instances are represented in computable form by one or more simple or complex spatial objects. Buildings are represented by points or polygons, rivers and roads by lines, lakes by polygons, and so on.

The same conceptual modeling process can be performed for any spatial system (a computer chip, a computer network, a protein structure, etc.) listed in the introduction. The reader is invited to perform a thought experiment using the outlined approach and to construct some spatial models for themselves.[5]

It may be surprising to learn that the conceptualization process leading to the representation of phenomena or entities (in our case a computer representation as an object) is not a creation of the modern science but one that it is older than we could possibly think. This is made clear in the following quotation:

> To make a particular object is to make this particular object out of general [material]. ... if we produce the form we must make it out of something else. Form means type. But we produce or beget a type out of particular material and what is then produced is an individual of this type ... [if we make a bronze ball out of bronze] ... the individual object is the bronze ball; but a type is sphere in general.

This quotation comes from *Metaphysics* by Aristotle (384–322 B.C.) as quoted by Loomis (1943). If, in this cited fragment of Aristotle's work, we substitute the spatial object for a bronze ball and the entity for the form, the fragment reads as the prescription for the creation of a computer representation of real-world phenomena. Of course, bronze in the example means obviously bits and bytes.

Spatial Data Structures

Before we proceed with further discussion of spatial evolutionary models and algorithms it is worth reviewing data structures used to represent spatial phenomena. Some readers may not need such a review: they are invited to skip to the next section. However, we think that for an equal or even a larger number of readers such a review, particularly at this point in our discussion, will provide a much needed basis from which to look at evolutionary models of spatial phenomena. People who are less familiar with spatial models may not be familiar with the specialized techniques or dedicated data representations employed by GISs. This review aims to show some of the qualitative differences that spatial data structures exhibit from general computational structures.

Spatial data structures are formalisms that allow entities and their relationships identified in a spatial model to be realized in a computational framework. Spatial data structures consist of a theoretical apparatus derived from topological and geometric theory and a set of computational methods to implement them. In the following sections the theory is first outlined as a guide to the generic approach to spatial data structure design, then secondly the computational structures are described, and finally the available operations are outlined.

Theoretical Apparatus

In history the roots of geometry are in surveying, especially the ancient Egyptian practice of agrimensa, or field measurement, after the Nile's annual floods. By contrast the study of topology means the study of form and derives from the mathematics of the ancient Greeks. In a sense, therefore, geometry began as the science of "experienced" concepts of shape and position, while topology was the science of all possible geometries. This was first explicitly recognized in Klein's famous "Erlangen" series of lectures on mathematics in 1872 when he argued that geometries could be derived from the study of topology. Hence, the fundamental theories on which spatial data structures are based derive ultimately from topology.

In Klein's conception, topology could be defined as the study of those geometric properties that remain invariant (unchanged) under topological transformations such as a stretching or folding. Such properties, for example, "connectivity," "adjacency" and "containment" can be termed topological properties. By contrast, distance, area, and bearings are not topological properties as they are always changed by a topological transformation. Maps of public transportation networks are frequently drawn in a stylized form in which the distances and

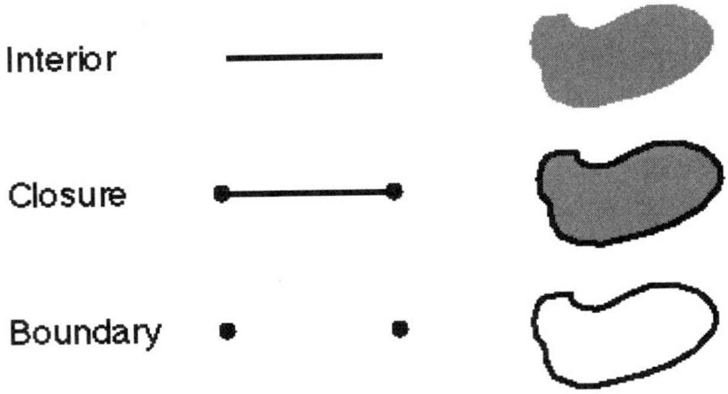

Figure 2-2 Definitions of topological terms.

shapes are arbitrarily transformed from the "real" values, while the topological properties remain unchanged. Probably the most famous example of such a "topologically transformed" map is the map of the London Underground.

"Pointset" topology based upon set theory is used to study topological properties. In pointset topology a theoretical "neighborhood" of any shape or dimension around an arbitrary point is made of an infinite number of points. If the points on the boundary are included in the neighborhood it can be modeled as a closed set and termed the "closure." If the points on the boundary are not included and the neighborhood is only made up of its "interior" then it can be modeled as an open set (figure 2-2). The real-world counterparts to these concepts might be a garden (interior), its wall (boundary), and the legal property (closure) if it included all of the garden and the wall. Pointset neighborhoods can be embedded in any dimension, although only a two-dimensional embedding is considered here. In a two-dimensional embedding such pointset neighborhood can be composed of one- and two-dimensional geometric forms.

These pointset neighborhoods can be transformed while retaining their topological properties if they are "connected." "Connected" neighborhoods can be defined as having an unbroken path from any point to any other point within the neighborhood. A "connected" neighborhood with no holes is referred to as "simply connected." Theoretical objects can be made from pointset neighborhoods; Worboys (1995) has suggested that such objects can be divided into points and 'extents' (see figure 2-3). The "extents" can be further subdivided into one-dimensional extents (boundaries with no interior) and two-dimensional extents (the full closure of the object). Note if the "boundary" of a one-dimensional neighborhood is non-self-crossing it can be termed "simple." The names of these precisely defined topological forms are deliberately generic to avoid confusion with the wide range of terms used in commercial GISs. Such theoretical terms are also of considerable use when designing data transfer standards for the exchange of data between different GISs. The U.S. Spatial Data Transfer Standard (SDTS) uses such terms.

Algebraic topology is the study of how such theoretical topological objects can be assembled into complex objects using discrete geometric primitives. The simplest approach to algebraic topology is to use "simplices" as primitives, that is, point (zero-simplex), straight-line segment (one-simplex), and triangle

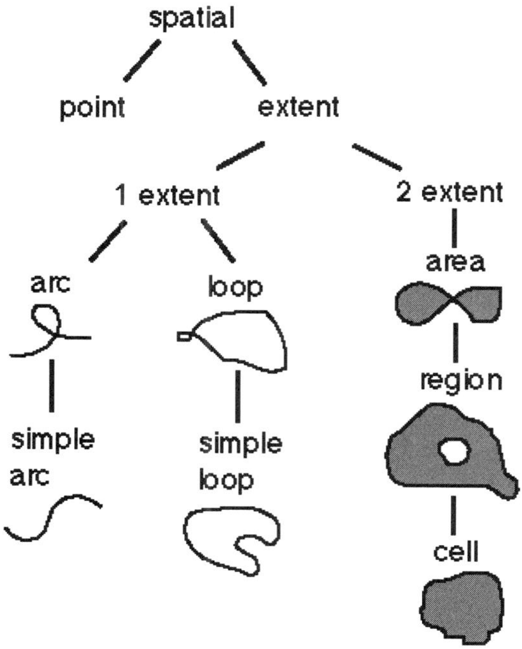

Figure 2-3 Geometric objects in theoretical pointset topology (after Worboys, 1995).

(two-simplex) forms. More sophisticated "cellular" approaches allow multiple segment lines and closed two extents ("cells") formed from multiple segment lines. These approaches to algebraic topology require that the pointset neighborhood be regularized so that "cells" are closed and connected to avoid any ambiguity. Graph theory can be used in the manipulation of sets of nodes (junction points) and edges (lines), which may form objects or trees. There is no notion of interior in a graph, which is only concerned with connectivity.

Each commercial GIS has selected from this set of theoretical structures to implement spatial data structures. In implementing a spatial evolutionary algorithm it has been necessary to choose a simple theoretical foundation to minimize the complexity of the algorithm. Commercial implementation of a spatial evolutionary algorithm would require the redesign of our framework to fit with the general theoretical framework employed in the system.

Computational Structures

Such theoretical topological objects need to be realized in a GIS. By embedding objects from discrete "cellular" algebraic topology in a two-dimensional plane with a fixed metric, geometric representations can be made. This is the approach that Euclid took in his *Elements*, written around 200 B.C. Pure Euclidean geometry (EG) defines points by the angle and distance that they are from an origin (the theoretical line from the origin to the point is known as the "vector"). One and two extents can be created by joining sets of such point "vectors." Operations

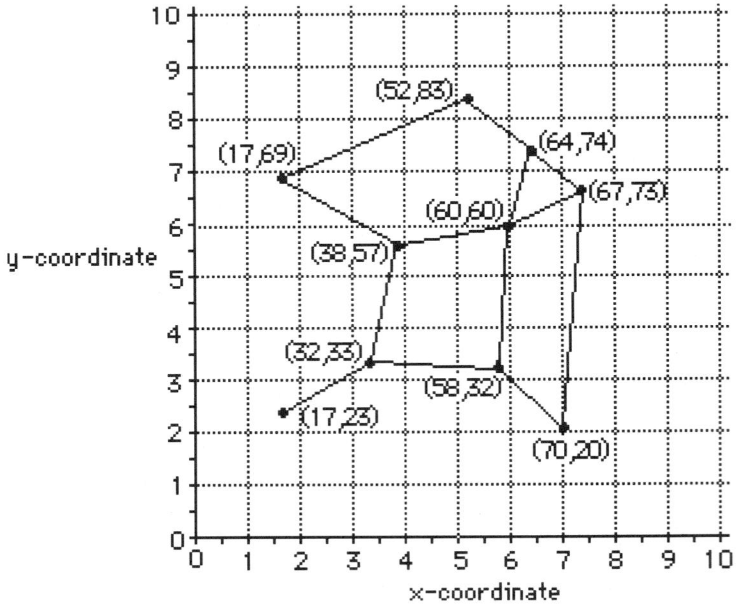

Figure 2-4 Coordinate Euclidean geometry.

can also be carried out on the geometric objects using trigonometric procedures, for example, the angles of arbitrarily shaped triangles can be calculated using the cosine rule and distances could be calculated by creating right-angled triangles and using Pythagoras' theorem. Note that curves defined by mathematical functions are not usually included in implementations of EG as they are functional in nature.

With the adoption of the two-dimensional Cartesian frame in the eighteenth century, EG was now reexpressed in coordinate form. This has subsequently facilitated the implementation of coordinated EG in finite-precision computers. In coordinate form a "vector" point can be represented as an x, y coordinate pair, to one extent by a straight line between such points (multiple-line-segment lines can be termed polylines), and a polygon by a polyline closed on itself (figure 2-4). GISs based on this system are implementations of this EG in a Cartesian coordinate framework where the framework is usually a "projection" of the Earth from three dimensions to two.

GISs are implemented by developing EG handling tools that create, edit, and query spatial objects and link them to attribute information stored in a database management system (DBMS). However, the performance of the tools to access the data is crucially dependent on the data structures. There are two key objectives, namely, efficient access to the large quantities of geometric data; and the provision of a spatial key into the data. Each GIS has its own methods and terminology for geometric data structuring, the most generic being a computer-aided drawing (CAD) approach where the points, polylines, and polygons are hierarchically indexed (figure 2-5). This kind of data structure can be queried

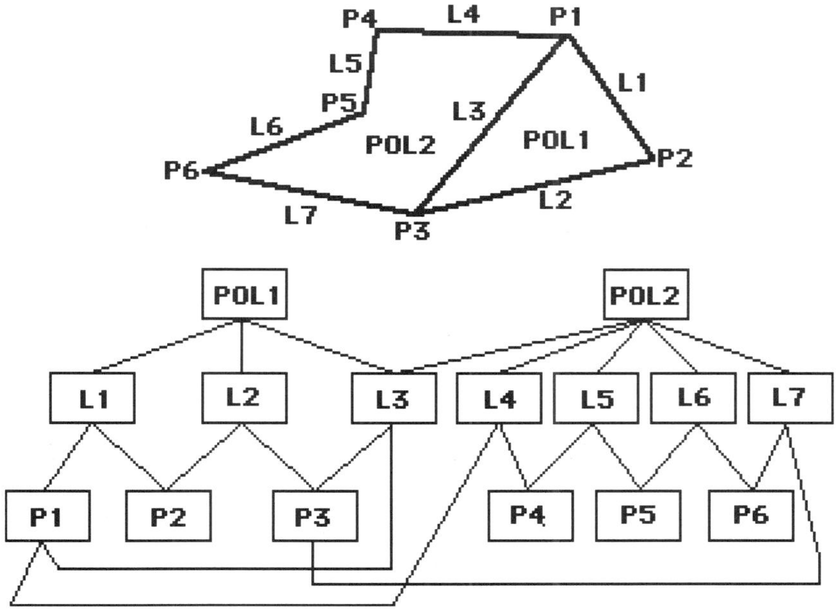

Figure 2-5 A CAD-type geometric data structure.

to establish topological relationships but does not store the relationships explicitly.

Providing a spatial key into the geometric data also involves the determination of topological relationships among the spatial objects, specifically adjacency, connectivity, and containment. In CAD data structures these relationships must be calculated at query time. However, specialized GIS data structures have been developed (e.g., POLYVRT in figure 2-6) that determine and store topological relationships automatically by entering the topological data in tables (Peucker & Chrisman, 1975). This approach assumes "planar enforcement" of the spatial objects so that all lines that cross intersect and polygons cannot overlap. Topological relationships can be established as a batch process or dynamically. In the process of developing a spatial model the spatial objects are usually divided into groups of objects that are permitted to intersect geometrically, for example, all objects with attribute "road". In a GIS such groups of geometric objects are stored in separate storage locations or "layers." GISs provide tools for the management of spatial objects within layers.

There is an alternative approach to the representation of space known as the "field" approach that is also widely used in GISs. A theoretical field is a measurable quantity that can be defined anywhere at an infinite number of points. As in the case of algebraic topology that is implemented using coordinated Euclidean geometry, the spatial variation of the field approach is made discrete using geometric tesselations (sets of connected discrete two-dimensional units). While regular tesselations can be formed from triangles, hexagons, and squares, only squares arranged in a grid are commonly used. A grid of squares ["cells" or

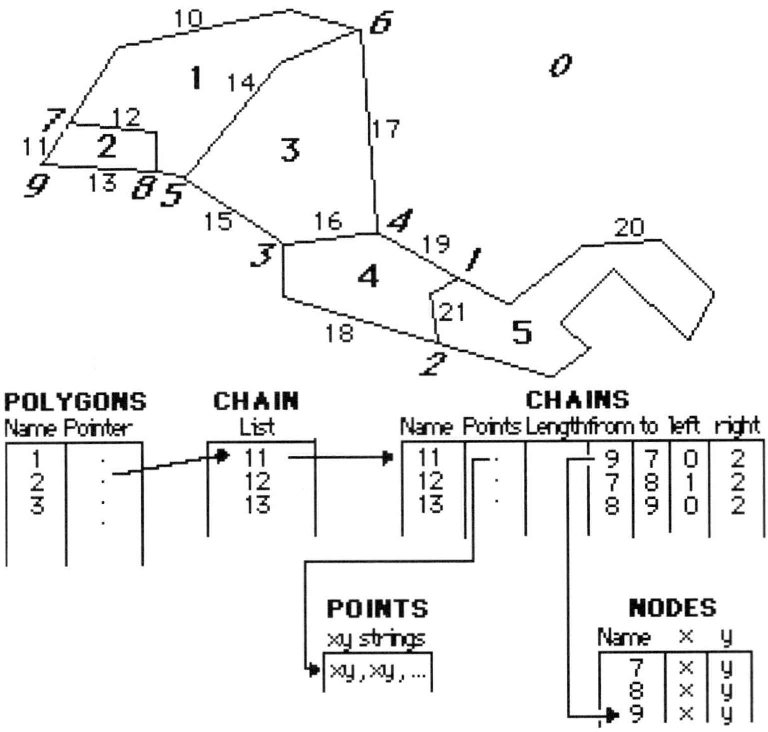

Figure 2-6 The POLYVRT topological data structure.

"pixels" (Picture elements)] is easy to handle in computers as memory arrays are logically similar to the physical imaging technology of faxes, scanners, satellite sensors, cameras, videos, and visual display units. Such imaging systems capture/display the reflectance of a real-world "scene" for each pixel in the "raster" within its range of sensitivity in the visible/nonvisible part of the electromagnetic spectrum. The "scene" can vary from the flat surface of a map to the surface of the Earth viewed orthogonally or obliquely from an aerial or terrestrial platform which itself may be static or moving.

A key characteristic of raster data is the spatial relationship of each pixel to the real-world observation, as, by definition, the stored value in the raster must apply homogenously within the pixel. When the systems capture *continuously varying data* by taking all the values obtained by the sensor within the area of the pixel they must then apply "pixel averaging" to get a mean value (e.g., satellite imaging and map scanning). By contrast, single measurements/samples at a point can be "assigned" to the whole area of a pixel even though in reality there may be (unsampled) internal variation within it (e.g., digital elevation models and risk factors). In both cases the boundaries of the pixel will become sharp discontinuities (if the values stored in each pixel are different) even if in reality the variation is smooth and continuous. When the systems capture *discretely varying data* each pixel must be assigned a value corresponding to the spatially dominant observation within the pixel. Some GISs offer the possibility to store point measurements as regular grids of points called "lattices" where the point values are not assigned to pixels by default but remain stored as points. This data structure offers the flexibility of dual display as either a surface with linear interpolation between the grid of points or as a "point-assigned" raster.

Clearly, the size of the pixel on the ground or the frequency of sampling of points defines the resolution of the data set in relation to the original spatial phenomena. Such pixel resolutions may be prefixed, as in the case of some Earth-observing imaging systems and some fixed-resolution sampling. Frequently, however, pixel resolution can be arbitrary. In some cases the underlying field being discretized in a raster can be nonisotropic, that is, there is a directional bias in the values. This is the case in travel time isochrones for calculating areas at equal travel time from some point.

In storage terms, raster data always has an implied spatial extent. By knowing the dimensions of the array, the real-world size of the pixels, and the coordinates of one corner of the raster, it is possible to find a coordinate position anywhere within the raster. Hence most raster data at a storage level consists of a header with dimensions, pixel size, and coordinates followed by the attribute data for the pixels as a stream in row order. When such data is stored in 80-column ASCII data exchange files a single row of data may take up many lines of the file. However, internally, the attribute data for the pixels is normally stored in binary form using binary digits (bits). The data range for the attributes is assigned to a range of binary numbers (4 bits = 16 alternatives; 8 bits = 256 alternatives; 24 bits = 16.8 million alternatives) and a legend relates the real values to the binary data (figure 2-7). Some GISs have tables to store the pixel attribute data pixels linked to row/column position.

Such rasters normally contain redundancy, since some values are repeated in the array when adjacent. Hence, it is necessary to structure and compress the raster to permit more efficient handling (Samet, 1990). This can be achieved by the use of "scan order" compressions using run-length encoding. In this technique a predetermined path is defined through the raster and only changes of value along the path are stored (figure 2-8). While many different scan orders have been proposed, there is no one ideal order as every raster has a different form of spatial variability.

An alternative form of compression is to recursively subdivide the raster (usually by four) until any quadrant is homogeneous when it is no longer subdivided. This compression is stored as an indexed tree known as a quadtree (figure 2-9). All these compression schemes are lossless, meaning no data is lost in compression, but lossy schemes are often used in computer graphics when the rasters only need to look good to the human eye. These principles of raster structuring can also be applied to large vector spatial databases to provide rapid spatial access.

Geometric Operations

Operations on spatial objects usually consist of two steps: a heuristic—a rule of thumb or simplification in order to reduce the search for solutions in a large

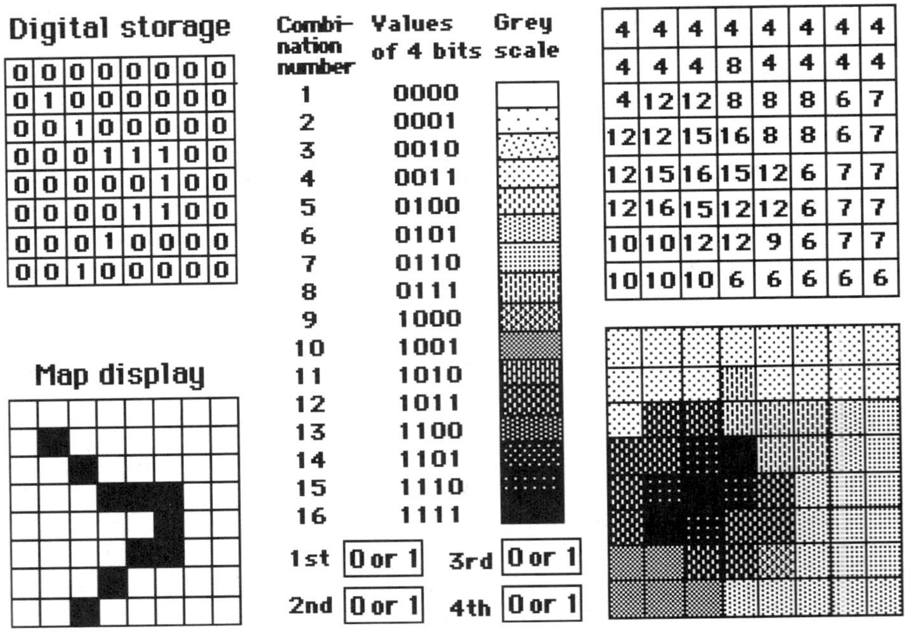

Figure 2-7 How rasters are represented by bits.

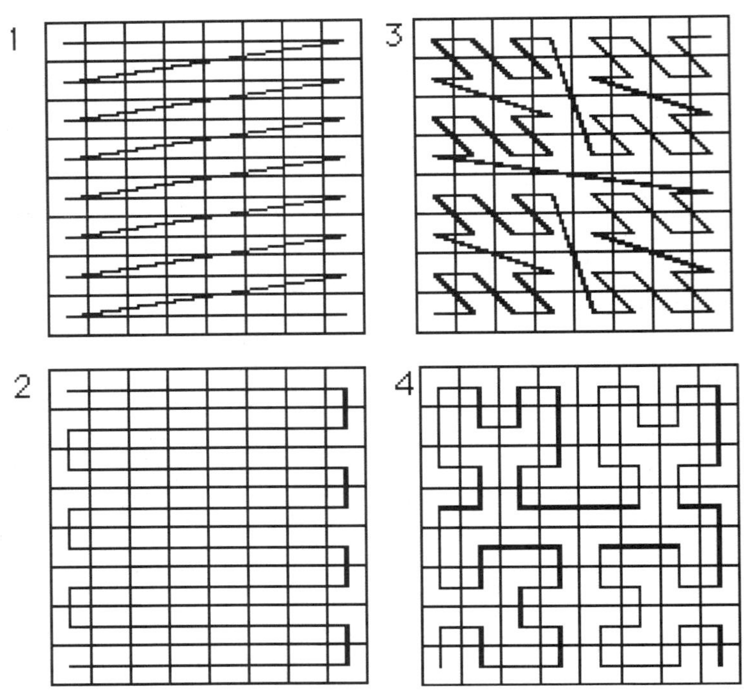

Figure 2-8 Scan orders: 1, row order; 2, row prime order; 3, Morton order; 4, Peano order.

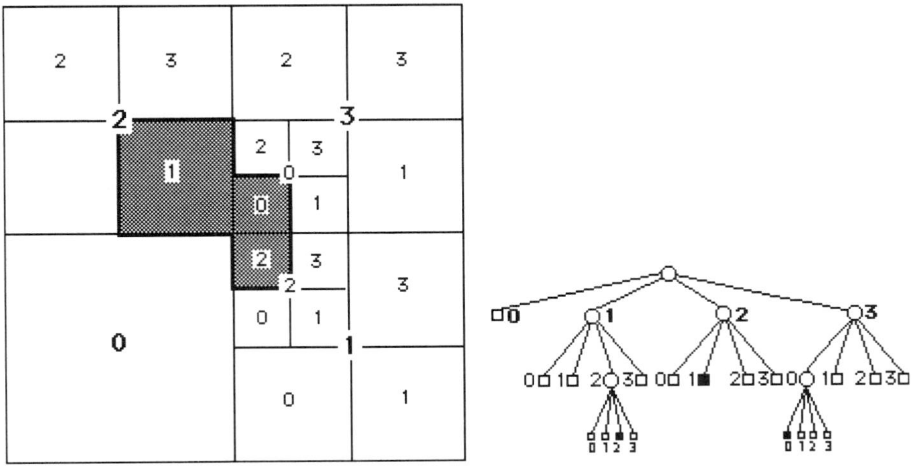

Figure 2-9 Quadtree structuring for the gray-shaded pixels reducing the number of stored pixels.

problem space; and an algorithm—a procedure consisting of a set of unambiguous rules that specify a solution to a problem. Both of these are required for effective design of geometric operations as "special cases" are often encountered in GISs. The computational time complexity of an operation often increases as a function of the data input, so better performance than linear time is usually the aim. Geometric operations are also limited by the ruling precision of the computer (the number of binary digits used to represent real numbers), which determines the number of significant digits after the decimal point. When very accurate calculations are required, for example, surveying computations, numerical precision is very important.

Operations on spatial objects implemented using vector techniques are based on procedures developed in computational geometry. Fundamental to finding the geometric properties of spatial objects is the calculation of Euclidean distance using Pythagoras' theorem of right-angled triangles. By using the coordinate grid as a surrogate, right-angled triangles can easily be formed with sides of known length. Angles between arbitrary points can be calculated by forming a triangle subtending the angle and then using Pythagoras to calculate the length of the triangle edges and the cosine rule (for non-right-angled triangles) to calculate internal angles.

Geometric operations can also be defined for polygons. The area of a polygon (if topologically "simple") can be found using the "drop trapezoid" method. The centroid of a polygon is also a geometric property, but one that also depends on the definitions employed: it can be the center of gravity; the center of a circumscribing or inscribing circle, or the peak value of a surface fitted to the vertices by attribute value. The geometric operation used to find the centroid is usually the average of the all the vertex coordinates. However, this can give the wrong answer if the polygon is long, thin, and curved.

Topological operations determine interrelationships between spatial objects. Finding line intersections involves deciding if polyline intersection is possible (Douglas, 1990). Point-in-polygon determination uses the "crossings" algorithm to determine whether a line dropped to the point from outside the polygon concerned intersects the polygon an odd (inside) or an even number of times (outside).

These operations function at two distinct levels: interface to data structure, and algorithm procedure. Generally, operations on quadtrees have the best performance for three reasons: firstly, quadtrees lead to significant compression of the data volume; secondly, quadtree-compressed raster data can be accessed very quickly by converting the quadtree leaf codes to row and column addresses; thirdly, in some operations the tree can be operated on rather than the raster itself. These data structure issues mean that a single algorithm may actually be implemented in a variety of ways, with implications for performance.

There are a huge range of potential raster operations depending on the type of raster attribute data and whether multiple bands of data are available (as in remote sensing operations). The most widely used classification of raster operations was made by Tomlin (1985, 1990) who introduced a structure and notation for the operations called map algebra. In map algebra a "region" is a set of connected pixels forming an area (although there may be multiple regions with the same attribute) and a zone is the set of all the cells in a raster with a given attribute value (whether connected or not). Note that in map algebra all operations produce a new raster and so sequences of operations carried out in a raster modeling exercise can generate large numbers of maps. While the map algebra is not defined in strict mathematical terms, it does provide useful concepts of a generic spatial functions and it is in this context that map algebra operators have been used in the spatial evolutionary algorithm.

Readers interested in taking the study of spatial data structures further should consult Herring (1991) or Molenaar (1998) for formal treatments of the subject.

Spatial Evolutionary Modeling: Past and Present

Armed with this rudimentary understanding of spatial models and data structures we may progress to the next question: What are spatial *evolutionary* models? Spatial evolutionary models are models of spatial systems that are expressed within the framework of evolutionary algorithms:

Real-world phenomena → Entities → Spatial objects → Evolutionary model

This process is very similar to the process describing the conceptualization of spatial phenomena that was discussed earlier. However, it has one extra step that involves mapping of elements of a spatial model (entities implemented as spatial objects) into equivalent concepts of an evolutionary framework.

In an evolutionary framework, spatial objects representing entities can be thought of as evolving units. Attributes of entities can be represented in chromosomes of evolving units. A collection of entities representing a spatial system may be interpreted as a population of evolving units. A metric characterizing a spatial system may be interpreted as a fitness of the population. A detailed example of such a mapping will be given later in this chapter.

Systematic scientific investigations of modeling with spatial evolutionary models have been scarce. Much of the existing research has been devoted to solutions of particular spatial problems rather than towards providing a systematic approach to the modeling of this class of phenomena. Some of the published solutions are described in the following section.

Lin et al. (1993) described a genetic algorithm for three-dimensional container packing. The objective of the container-packing problem is to pack, in layers, a maximum number of smaller boxes (*parallelepipeds*) into a larger one in order to minimize unused space. The solution is subject to restrictions on the weight and shapes of parallelepipeds. The genetic algorithm uses an individual with multiple chromosomes. Each chromosome represents one layer of boxes and each individual represents one layout of boxes composed of several layers. The genetic algorithm operators reflect the design of chromosomes: a partially mapped cross-over that has been designed to preserve the structure of layers and mutation implemented as shuffling of genes within a layer. Lin et al. compared the performance of their genetic algorithm with the heuristic SMILE (Lin et al., 1993), which is believed to provide the best solution for this type of packing problem. They reported that in 99% of tests their GA performed better than the SMILE.

Pargas & Jain (1993) proposed the use of multidimensional chromosomes for modeling of a two-dimensional (2-D) bin-packing problem. The objective of the 2-D bin-packing problem is to pack a finite number of 2-D objects (squares or rectangles) into a 2-D bin of a given shape to maximize the number of packed objects. Pargas & Jain's algorithm is designed to pack polygons of a unit length that loosely resemble block letters: C, E, F, H, I, L, and T. A set of polygons, representing the solution to Pargas and Jain's problem, is coded as a set of tuples. Each tuple has two indices; one representing a type of polygon and one specifying an angle of rotation of the polygon. Only four cardinal angles are allowed (0°, 90°, 180°, and 270°). A chromosome also contains the total length of the solution (a number of polygon tuples).

The operators implemented by Pargas & Jain (1993) only *resemble* classical evolutionary operators. The selection operator has been implemented as the selection of the best individual from an existing population—provided the individuals in its neighborhood were at least as good. The mutation (termed perturbation in Pargas & Jain's model) is generating a random solution. Pargas & Jain (1993) reported that the algorithm achieved the goal of 80 percent bin utilization in all of the tested cases.

A multidimensional chromosome structure was also proposed by Al-Attar (1994). Al-Attar's multidimensional chromosome design was used in modeling of a scheduling problem. Apart from researching complex chromosome structures, this research has no specific significance for spatial modeling. However, Al-Attar provided several general recommendations for the use of multidimensional chromosomes that are considered to have broad applicability. Al-Attar considered it to be easier to represent real-world problems with multidimensional chromosomes. He acknowledged that the use of multidimensional chromosomes requires the design of specific operators. In addition, he pointed out that these operators should work on each "dimension" of a chromosome separately and should produce results that are "legal" (i.e., within the chromosome domain).

Juliff's pallet-loading problem (Juliff, 1994) might be regarded as a reasonably comprehensive study of a genetic algorithm with multidimensional chromosomes. The objective of the pallet-loading problem is to pack cartons into layers, and layers into pallets. Such packing must be subject to the specific geometric characteristics of cartons, layers, and pallets, as well as to the physical properties (weight) that restrict the numbers of layers on a pallet. Individuals in Juliff's algorithms are "loads." Each load is characterized by the number of layers and the number, type, and order of pallets. Information about the load is coded in three chromosomes per individual. The information about the number of layers and their order is contained within the first chromosome. The information about the type of pallets per load is contained within the second chromosome. The sequence of pallets is coded within the third chromosome. For his problem, Juliff designed multidimensional chromosomes and a suite of operators (selection, cross-over, and mutation). He stated that:

> [R]epresentation of the multidimensional problem with a single chromosome requires highly specialized and customized operators. Thus, by using the multidimensional chromosome structure—with each chromosome representing a single future—one achieves the simplicity (and the clarity) of the design as well as of operators (Juliff, 1994).

He reported that the multidimensional chromosome GA performed much better with the test problems that similar single chromosome designs.

A unique innovative approach to the study of spatial modeling was reported by Delahaye et al. (1994). The authors developed an evolutionary algorithm for partitioning of the control air space into exclusive regions called flight sectors. Each flight sector is subject to the rules of flight control. Flight sectors are created by tessellating the control air space into convex polygons. Each polygon (flight sector) is represented by its center (centroid) X and Y coordinates. A set of X and Y coordinates for all sectors of one partition constitutes a chromosome or an individual. A population of individuals represents alternative partitions of the same control air space.

Delahaye's evolutionary algorithm uses two operators: cross-over and mutation. Cross-over between sectors of selected individuals (an individual represented one partitioning of the control air space) is implemented as the arithmetic average of X and Y coordinates of two selected sectors. Mutation is implemented as a random change to X and Y coordinates of the center of a flight sector. Delahaye et al. (1994) reported that his algorithm performed much better than other methods designed for the same type of problems.

Several innovative applications of evolutionary methods in the field of geographic information systems have been reported. Cooley & Hobbs (1992) used a bit-encoded GA for the determination of optimal class partitioning of continuous data for display. Dibble & Densham (1993) discussed potential applications of the GA to the generation of alternatives in spatial decision support systems (SDSS). They suggested that the GA could be very useful in formulating several alternative solutions to modeling (Pareto solutions). A similar application of genetic algorithms has been reported by Bennett et al. (1996). The authors described the use of the GA for the multiobjective analysis of land-use scenarios. An interesting feature of their application was genetic encoding of the rasterized land-use data using the Morton linearization scheme. Armstrong & Bennett (1990) described the use of evolutionary algorithms for knowledge extraction from the contaminant database. Similarly, using Holland's (1993) GA classifier concept, Dibble (1994) proposed the use of the GA as a classifier for processing of spatial information. Keller (1995a, b) proposed the application of the GA to line generalization.

Hosage & Goodchild (1986) described the GA for the *p-median* problem. Their design was based on a canonical GA, both in coding of genetic material and in the form of operators. The authors reported that their algorithm produced less satisfactory experimental results than those obtained with known p-median algorithms. However, this result might have been anticipated as already existing specialized algorithms for well-defined problems such as the p-median problem, will, in most cases, outperform a weak optimization method[6] such as the GA.[7] This is particularly true if the GA is not hybridized or made problem specific.

In an unpublished experiment, Bianchi & Church (1992) designed a GA for a location problem (p-median) similar to that earlier studied by Hosage & Goodchild (1986). Unlike Hosage & Goodchild (1986), they used integer coding and the mutation operator. The authors reported that their integer-coded GA performed better than Teitz & Bart's (1968) algorithm in all tested cases.[8]

Several novel applications of evolutionary algorithms to spatial modeling have been recently proposed by Hobbs. In one of the earliest of these applications (Hobbs, 1993), a canonical GA was used for the discovery of knowledge in spatial databases. In this design, a genetic algorithm performed a supervised classification of spatial data.

In his subsequent application, Hobbs (1994) proposed the use of evolutionary methods for the definition of catchment areas. His research focused on modeling of contiguous areas associated with banking branches of a building society. The catchment area for each branch (the area generating revenue of the branch) is composed of several postal sectors.[9] Each branch shares revenue from several postal sectors. A chromosome (individual) represents one particular assignment of postal sectors to branches. Evolutionary operators (selection and cross-over) are designed to account for the representation of postal sectors and for their topology. Although Hobbs (1994) reported encouraging results from the tests of his GA, he cautioned that the research was still in its early stages and no ultimate conclusions should be drawn regarding the usefulness of the proposed evolutionary method for the modeling of catchment areas.

In another of his experiments, Hobbs (1995) reported an application of a GA to spatial clustering of street networks. The objective of Hobbs' clustering problem was to aggregate several houses into the network areas, taking into account the topology and contiguity of networks. To solve this problem, he proposed a genetic algorithm in which the elements of the network were represented as genes. Genetic operators (selection, cross-over, and mutation) operated on the networks, preserving their connectivity. In his GA, an individual was defined as a particular "partition" of the network. The findings were encouraging, but only preliminary results were available when published.

A genetic algorithm for site selection analysis, first reported by Pereira et al. (1966), has arguably been the most complete application of a GA to spatial modeling, to date. Pereira designed a GA for the generation of location alternatives, given a set of location criteria. Pereira's algorithm was a bit-encoded canonical GA with a fitness function derived from a spatial overlay and with spatial data integrated into a genetic algorithm in the form of a layered model. In addition to the detailed description of the design of his GA, Pereira reported the type of statistics used to monitor GA performance (on-line and off-line performance) as well as ranges of values for setting of genetic operators and their parameters. Pereira's algorithm represented a successful integration of spatial data into a spatial evolutionary model. In fact, this approach, if properly generalized, could serve as a simple framework for evolutionary modeling of spatial problems.

Recently, Brooks (1996) reported the design of an evolutionary algorithm for region-growing formulas. His algorithm was designed to determine clusters of specific shape and size, given selective clustering criteria and the locations of cluster centers. Brooks' genetic algorithm encoded the parameters of region-growing formulas. Evolutionary operators (cross-over, mutation, and creep) were designed to operate on these parameters, preserving their integrity. Brooks' algorithm was integrated with a GIS via region-growing formulas. In test trials, Brooks' GA outperformed a region-growing heuristic that was used for comparison.

Openshaw (1988, 1992, 1995) was first to predict the advent of, and the need for, adaptive modeling methods in spatial processing. Moreover, he went on to design successful applications using these methods. In 1988, he proposed the automated modeling system (AMS), employing an evolutionary algorithm as one of the methods for automated selection of the parameters of spatial interaction formulas. In 1992, he suggested the use of evolutionary methods in GIS modeling. These methods provided an attractive alternative to traditional spatial models. Recently, he and his colleagues (Openshaw, 1995, Turton et al., 1997; Openshaw & Perree, 1996) presented a method based on Koza's (1991) genetic programming for the derivation of spatial interaction models, reporting very satisfactory results from the experimental trials. Table 2-2 presents a summary of research on spatial evolutionary models.

- Past research has been dominated by implementations of evolutionary models cast into the syntax of evolutionary algorithms—a modeling approach that has been known to produce inferior models for complex problem domains.
- No general framework or guidelines for the evolutionary modeling of spatial problems has been formulated or even proposed; such a framework would facilitate the design of new models, operators, and representation schemes.
- There has been no attempt to design hybrid evolutionary algorithms for spatial models. Given the observation that problem-specific modeling algorithms are central to GIS technology, future research into evolutionary methods that integrates spatial modeling algorithms might well result in the emergence of a new generation of enhanced modeling methods.
- There has been no research into the properties of evolutionary operators of spatial models. Consequently, there is no understanding as to which operators perform well, under what conditions, and what settings of operators should be

Table 2-2 Summary of research on evolutionary methods and spatial information processing

Class of Application	Type of Problem	Reference	Characteristics of Proposed Algorithm
Multidimensional chromosomes	Bin packing	Pargas & Jain (1993)	Problem specific coding, coding specific operators
	Pallet loading	Juiff (1994)	Problem specific coding, coding specific operators
	Container packing	Lin et al. (1993)	Problem specific coding, coding specific operators
	Scheduling	Al-Attar (1994)	Problem specific coding, coding specific operators
Location problem	Air space partitioning	Delahaye et al. (1994)	Real value coded genes representing locations in 2-D, space processed using tessellation algorithm
	p-Median	Hosage & Goodchild (1986)	Canonical genetic algorithm
	p-Median	Bianchi & Church (1992)	Integer-coded genetic algorithm, problem-specific operators
Spatial interaction models		Openshaw (1988, 1992, 1995)	Genetic programming
Suitability modeling		Pereira et al. (1996)	Canonical GA integrated with GIS
		Brooks (1996)	Hybrid GA integrated with GIS
Aggregation		Hobbs (1994, 1995)	Problem-specific coding, and problem-specific operators
Knowledge discovery		Hobbs (1995)	Canonical GA for clustering
	Classification	Armstrong & Bennett (1990)	GA classifier
	Data mining	Dibble (1994)	GA classifier
Multiobjective optimization	SDSS	Dibble & Densham (1993)	
	Multiobjective optimization	Bennet et al. (1996)	Use of Morton order for the encoding of the spatial data in chromosomes
Generalization of spatial data	Line generalization	Keller (1995a, b)	
Display	Classification of continuous data	Cooley & Hobbs (1992)	Canonical GA

used. Without this knowledge, the development of new models will have to be based upon laborious experimentation and, hopefully, good luck—not a very reliable modeling approach.
- Lastly, the striking similarity between data structures and data structure manipulation algorithms in GP and spatial information systems has not been explored at all, to the detriment of both fields.

We believe that there are many problems for which evolutionary modeling of spatial problems can be applied. The goal of future research should be to provide the framework for modeling of spatial problems with evolutionary algorithms. Such a framework should contain guidelines for the design of genetic representation of spatial phenomena and for the development of spatial evolutionary operators. We hope that some of the concepts discussed in this book will prove to be of help in this endeavor.

Spatial Evolutionary Model—an Overview

As we pointed out in the last chapter, there are many evolutionary models of spatial phenomena, with each model designed for one particular problem. Clearly, what is needed is not yet another model but the framework for modeling of spatial problems with evolutionary algorithms—something like the spatial evolutionary model we have proposed. In this and subsequent chapters, we outline and elaborate such a model. In order to demonstrate its validity and to clarify any ambiguities associated with the abstract definitions we will discuss our spatial evolutionary model in a context of a spatial model of a wireless communication system. The model of the wireless communication system represents a class of covering problems. However, we hope that the reader will see this model in a much broader context as it has elements common to a broad variety of spatial phenomena.

The general formulation of the covering problem involves the determination of the minimum number of facilities to cover the demand and is defined by Church (1984) as follows:

$$\min Z = \sum_{j=1}^{n} c_j x_j$$

such that

$$\sum_{j=1}^{n} a_{ij} x_j \geq 1$$

$$i = 1, \ldots m$$

$$j = 1, \ldots, n$$

$$x_j = (0, 1)$$

where c_j is the cost of assigning facility to a site j; x_j equals one if the facility is assigned to the site j and is zero otherwise; a_{ij} takes on the value of one if a

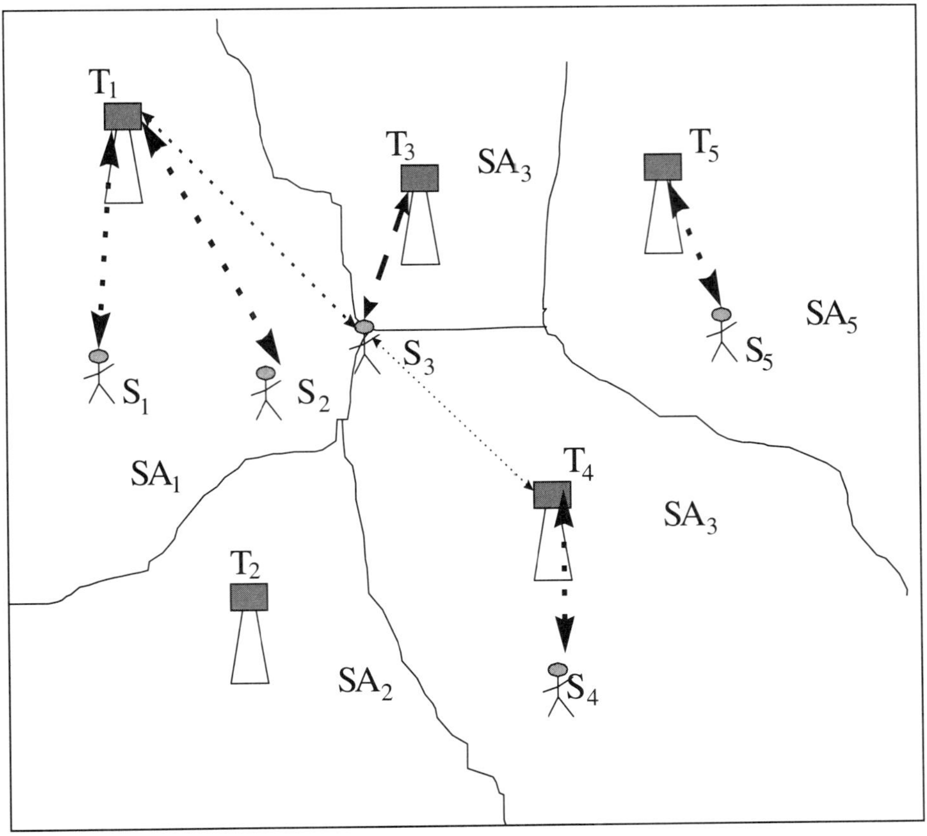

(T_i - transmitters, S_i - clients with receivers, SA_i - service areas of T. Dotted lines symbolize the communication links between the clients and transmitters)

Figure 2-10 A conceptual model of the wireless communication system.

demand at site *i* is covered by site x_j and the value of zero if it is not; *m* is the number of demand points; *n* is the number of facilities.

Covering problems touch upon diverse aspects of planning, such as the location of plants, public schools, police stations, libraries, hospitals, public buildings, post offices, parks, military bases, radar installations, branch banks, shopping centers, waste disposal facilities (Frank & White, 1974), radar stations (Bailey, 1992), and several other types of facilities (Bianchi & Church, 1992; Church, 1984).

In most general terms, the wireless communication system consists of a network of transmitters (T_i), service areas (SA_i), and clients using the services of the wireless system (S_i) (see figure 2-10).

Transmitters are connected to a switching station, which relays calls to the telephone network (POTS). Each transmitter (T_i) in the network operates over its service area (SA_i). Any client in the service area of a transmitter can communicate with this transmitter. At locations where two or more service areas overlap a client may receive a signal from more than one transmitter. The extent and shape of the service area of each transmitter are determined primarily by the power and antenna type of the transmitter and by the topography and land cover of the area surrounding the transmitter. Each transmitter has a capacity to handle a certain amount of traffic (demand). This capacity is expressed as a number of receivers with which a transmitter can communicate simultaneously. The set of transmitters forms a wireless communication system.

A wireless communication system can be characterized by several physical variables or metrics: the number of transmitters, the extent of the area over which transmitters receive and transmit a radio signal, signal strength (transmitted or received), and distribution of the demand.

Demand is a complex geographic variable that expresses the number of customers that potentially may use a wireless service. Demand is determined by the transportation patterns of commuters, shoppers, and vehicular traffic and, indirectly, by the distribution of retail shopping, business centers, and industrial and commercial parks. Demand is expressed in units called CCS[10] and erlangs.[11] The measures *total covered demand* and *total covered area* may be regarded as the summary measures of demand. They are functionally equivalent measures of the same system characteristics. This is particularly true whenever the demand for service is uniformly distributed across an area serviced by the wireless system.[12]

The wireless communication system is also characterized by a more subjective metric—*average signal quality (ASQ)*. The ASQ is a combination of an average signal strength and a signal interference (roughly equivalent to the percent of overlapping service areas). The average signal quality is a measure reflecting the "quality" of the transmission that the client of the wireless communication system may experience while using the system services.

The quality of the wireless communication system depends, to a large extent, on the location of transmitters. Transmitters located in favorable locations will assure desirable signal quality and high quality of service. Conversely, poorly located transmitters will not provide an adequate signal coverage, degrading the overall network performance (Boucher, 1992; Gamst et al., 1985; Lee, 1985; Brocken & Strelder, 1990; Chan, 1991).

The task of finding locations for transmitters that would assure the required signal quality and adequate demand coverage, given a number of transmitters of limited capacity and range, has always been a challenge and may be regarded as being more of an art than strictly engineering. The main objective of this task is simple: to locate a number of transmitters in a way that will provide acceptable average signal quality over the most of the serviced area. However, to reach this objective one has to simultaneously account for the distribution of complex traffic patterns (demand for the wireless service), the presence of geographic features (topography, morphology, land cover), as well as for the technical characteristics of the network (Boucher, 1992; Gamst et al., 1985; Lee, 1985; Brocken & Strelder, 1990; Lopez and Vlahodimitropulos, 1995; Kurner et al., 1993; Ohlson, 1995).

With so many diverse factors influencing the location, it is of no surprise that no comprehensive, analytical model of transmitter locations has been proposed. Yet, this problem is a perfect one for modeling using evolutionary methods.

Table 2-3 The wireless communication system within the spatial systems framework

Spatial Systems Features	Wireless Communication System
Spatial systems are composed of spatial objects	Transmitters and service areas are spatial objects
Spatial objects have designated properties	Transmitters and service areas have properties of location, power, capacity, demand, and shape
Spatial objects are inter-related	The set of transmitters forms the pattern dictated by the assigned communication channels, land cover type and terrain morphology
The system of objects has designated characteristics that allow for differentiation between several similar systems	The wireless communication system has assigned a metric that represents the quality of the wireless system as a whole. This metric can be the subjective quality of transmission, the average interference, or the average signal level, or capacity

How do we describe the wireless communication system within a four-point spatial system framework? In the context of our spatial framework, transmitters are defined as *spatial objects*, with the properties of *location, transmit power*, or *capacity*. The set of transmitters forms a definite pattern that is dictated by the assigned *communication channels, land cover type*, and *terrain morphology*. The area around transmitters is characterized (at each point) by the *signal strength, terrain morphology*, and *terrain relief*. Finally, the wireless communication system has a metric that represents the quality of the wireless system as a whole. This metric can be the subjective *quality of transmission, the average interference*, or *the average signal strength*. Table 2-3 summarizes these concepts.

Spatial Evolutionary Model—Formulation

We will assume at this point that the reader is already familiar with evolutionary terms such as *chromosome, genes, population, fitness*, and so forth, as explanations of these terms have been presented in the first part of the book.

The development of the spatial evolutionary models proceeds in three phases: (1) formulation of the spatial model; (2) representation of the spatial model in the evolutionary framework; and (3) definition of evolutionary operators.

In Phase 1, we select elements of real-world phenomena that are pertinent to the model and represent them as spatial entities and spatial objects. This phase follows the conceptualization process of spatial phenomena explained in the introduction. In Phase 2, we construct the evolutionary representation of the spatial model. In this phase, spatial objects formulated in the first phase are mapped into the evolutionary framework, that is, into *populations, organisms, chromosomes*, and so forth.

A schematic representation of these two phases is given in figure 2-11. In the figure, the real-world phenomenon—a wireless communication system—is represented in the form of a map, with ovals representing transmitters. Elements of the

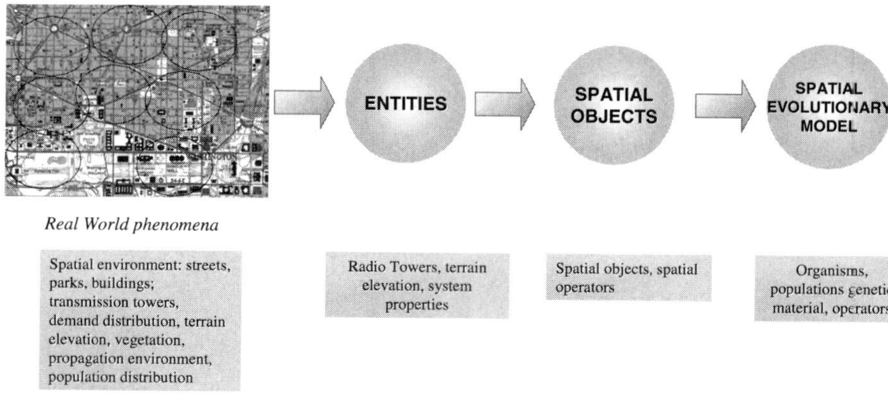

Figure 2-11 A conceptual representation of the steps in the spatial evolutionary modeling.

phenomenon selected for the model (entities) are listed in the box below the "entities" oval. Entities are represented as spatial objects and instances of spatial objects are expressed as a population of individuals containing chromosomes.

A population of individuals is represented in figure 2-12. A population is composed of organisms (small ovals inside the square). Each individual has three chromosomes (genes) [X, Y, R]. Each population has an index P and each individual an index i. A collection of P constitutes a hyperpopulation (not presented in the figure).

We are ready now to discuss Phase 3 of our modeling process that involves the definition of operators. The proposed evolutionary model has seven operators:

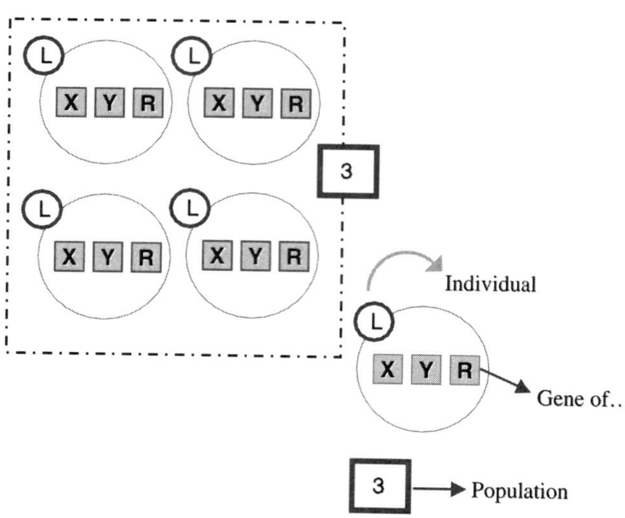

Figure 2-12 Symbolic representation of elements of a spatial evolutionary model discussed in the text.

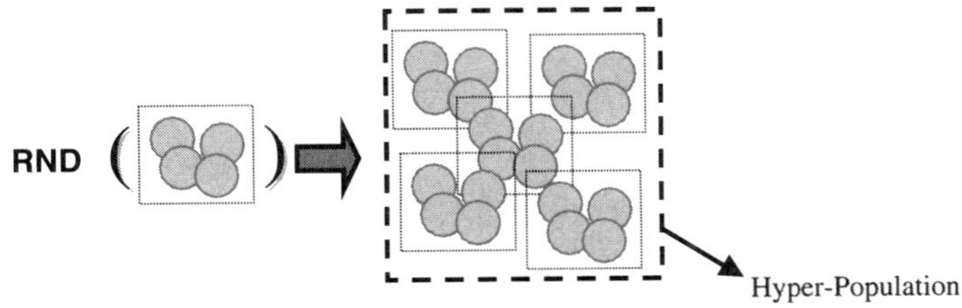

Figure 2-13 A symbolic representation of an initialization operator of a spatial evolutionary model.

- An initialization operator generates an initial hyperpopulation. The operator starts with the definition of the domain and generates [symbolized by $f(rnd)$] a number of populations. Operation of the initialization operator is presented in figure 2-13.
- A fitness operator measures the quality of populations with respect to the model objectives. A fitness operator takes a current population with its organisms and calculates their fitness according to the problem-defined objectives. Figure 2-14 represents a simplified fitness calculation.
- A selection operator selects populations (and organisms in these populations) with higher than average fitness. Selected populations are used to produce new populations in the reproduction (cross-over) operation. Figure 2-15 represents schematically the process of selection.

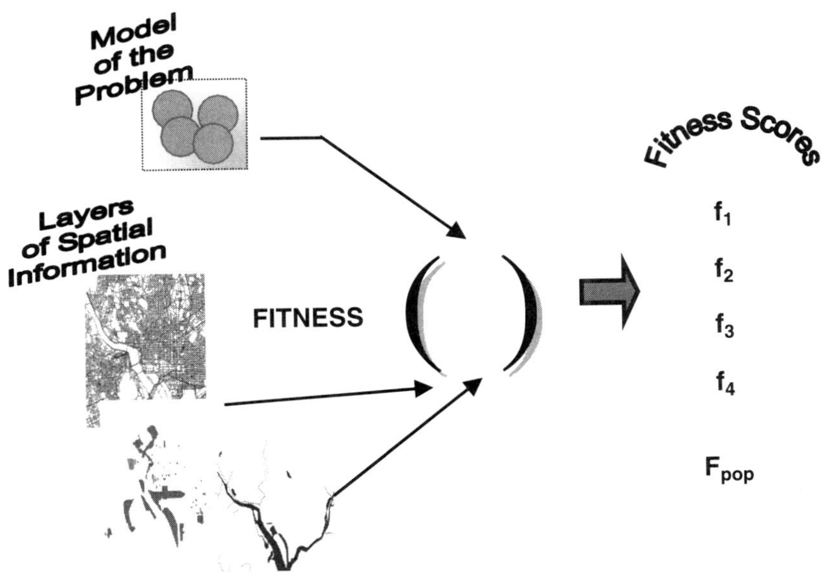

Symbols on the left of the FITNESS () function represent elements that will be included in the fitness calculation: a population, and spatial information in the form of separate thematic layers (see text for detailed explanation of this operator). Symbols on the right represent the fitness of each organism (f_i), and the fitness of a population of organisms (F_{pop}).

Figure 2-14 A symbolic representation of a fitness operator of a spatial evolutionary model.

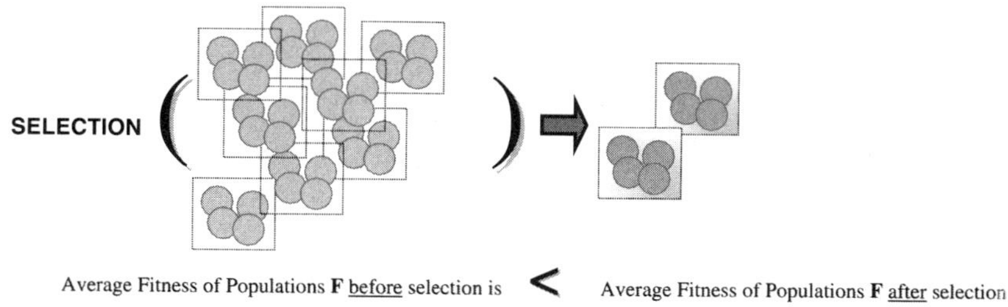

In this operation selected populations (on the right of the *Selection ()* function) have better than an average fitness

Figure 2-15 Symbolization of the selection operator.

- A mutation operator randomly changes the genetic material of organisms. This change may happen in only one organism or in all the organisms of the population. Mutation is illustrated in figure 2-16.
- A cross-over and mating operator recombines populations and generates new offspring populations. Each of the new organisms in the offspring population has chromosomes that are combinations of chromosomes of two input populations (parent populations). The crossover of two populations is illustrated in figure 2-17.
- A learning operator improves the fitness of organisms in a population (and a fitness of a population) between the evolutionary cycles.[13] The algorithm performs search over the problem space trying to locate more optimal configurations for the modeled system. Figure 2-18 shows how this algorithm works. The "search path" over the problem space for three organisms is illustrated by the dotted line.
- An objective function operator expresses the fitness of the evolutionary units (i.e., populations) with respect to the problem objectives. The fitness of the population is equivalent to the value of the objective function.

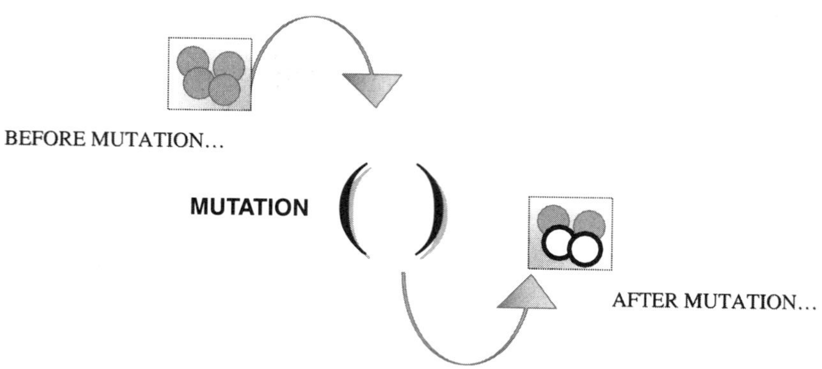

The shaded outlined circles (organisms) represent organisms that have mutated. The mutation shown affects only selected organisms in a population, not the whole population.

Figure 2-16 Representation of the principles of a mutation operator in a spatial evolutionary model.

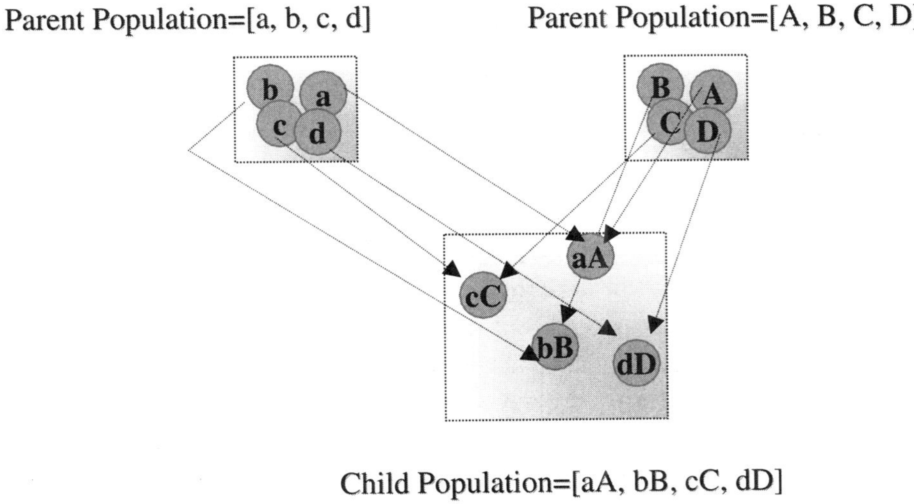

Figure 2-17 Representation of the cross-over operation on populations and organisms in a spatial evolutionary model.

The complete pseudocode of the spatial evolutionary algorithm is presented in figure 2-19. The bulk of work of the algorithm is carried out in a loop (lines 3–9). A generation is one cycle through this loop. The algorithm stops when the condition specified by the statement in line 9 is met. This conditional statement monitors the convergence of fitness scores of evolving populations.

Experimental Results

The capabilities of the algorithm were tested on three different models of wireless communication systems. Each wireless communication system had the same

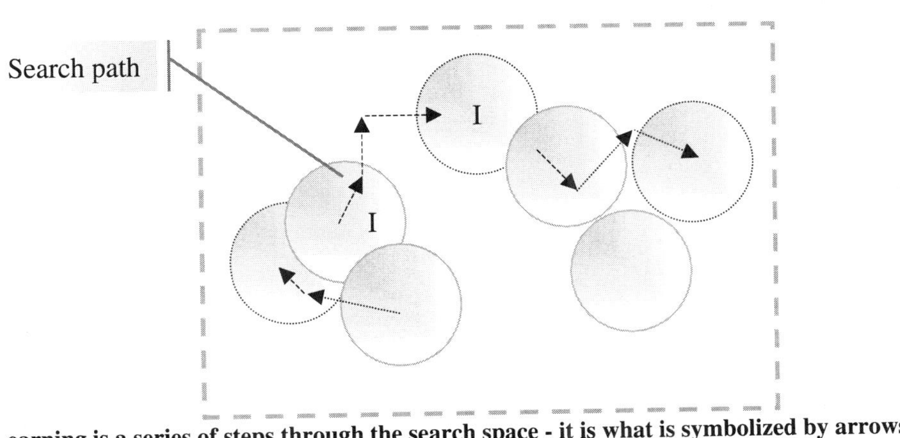

Figure 2-18 Representation of learning as implemented for the evolutionary model.

```
Line [1:] Initialize Population
Line [2:] Calculate Fitness
Line [3:] Loop:
Line [4:]     selection operator
Line [5:]     mate and cross-over
              operators
Line [6:]     mutation operator
Line [7:]     learning operator
Line [8:]     fitness operator
Line[9:]      Stop Condition
Line[10:] End Loop
Line[10:] End Run
```

Figure 2-19 The complete spatial evolutionary algorithm. Line number refers to the description in text. The real evolutionary algorithm is more complex than the one in this figure yet it retains the same framework.

number of transmitters (30) but a different size of service area (as expressed by the length of its radius). Each wireless communication system was associated with a different geographic data set. The data sets represented outlines of three geographic objects: a county (Suffolk County, N.Y.), and two groups of islands—Antilles, and Great Britain and Ireland. These three geographic data sets differed in the complexity of their shape and topology (connectivity). The data set for the county was a simple connected surface; the two other data sets were non-simply connected surfaces. Complexity of shape and topology of spatial data is what makes any spatial problem much too difficult for classical modeling methods and it is what makes spatial evolutionary algorithms—capable of handling such complexities—so attractive as a modeling tool. Each geographic data set contained a uniformly distributed demand. A demand greater than 0.0 was distributed within the object boundaries. A demand of 0.0 was assigned outside of the object boundaries but within the object bounding rectangle.

The results of the experiments are presented in figures 2-20, 2-21, and 2-22. Figure 2-20 illustrates the configuration of the wireless communication system as produced by the spatial evolutionary algorithm for the first data set. Figures 2-21 and 2-22 illustrate the configurations of wireless communication systems for the other two data sets.

Upon visual inspection, all three tested models show acceptable distribution of the transmitters over their respective geographic areas. In all three models, transmitters have been located almost without an overlap of serving areas and with very good correlation with geometry and topology of the spatial objects. In addition, for each transmitter, on average, 80 percent of its capacity was allocated. A minimal overlap between service areas and high capacity utilization

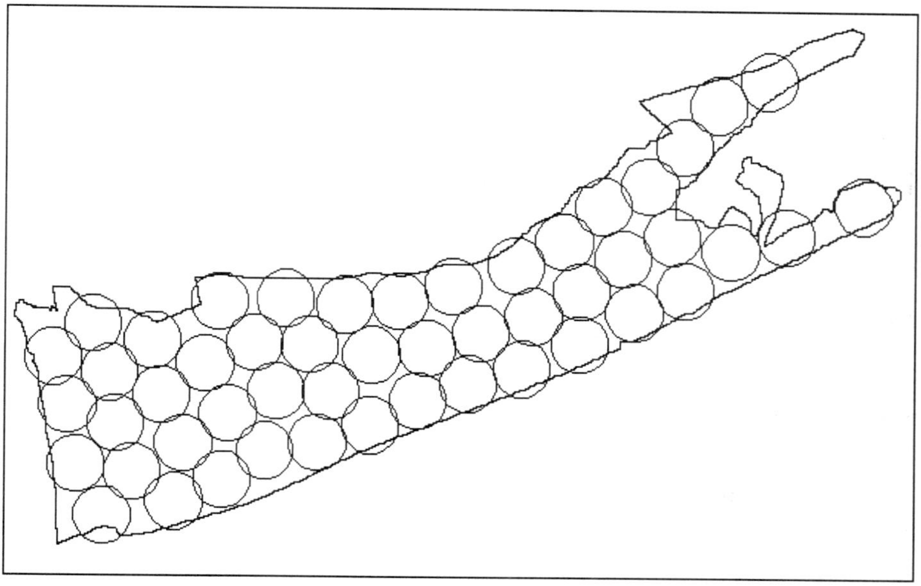

Figure 2-20 Representation of Suffolk County, N.Y. with a set of transmitters (circles) allocated using the spatial evolutionary model described. The demand in this example was allocated to the whole plane defined by the square (bounding rectangle) surrounding the outline of the county: the area within the county outline had allocated a demand "1" (demand present), and the rest of the area "0" (no demand).

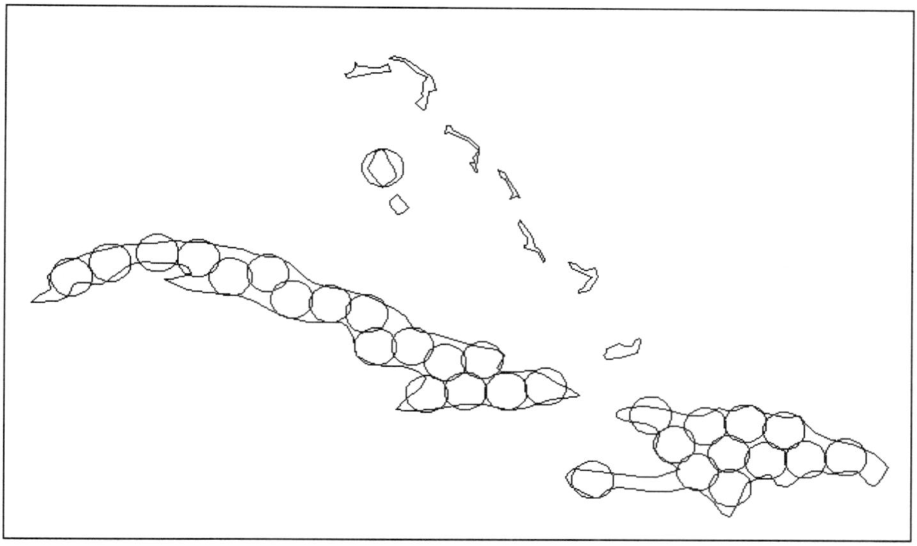

Figure 2-21 Representation of Antilles with transmitters (circles) allocated using the spatial evolutionary model described. The demand in this example was allocated to the whole plane defined by the square (bounding rectangle) surrounding the outline islands: the land within islands had allocated a demand "1" (demand present), and the rest of the area "0" (no demand).

Figure 2-22 Representation of the British Isles with transmitters (circles) allocated using the spatial evolutionary model described. The demand in this example was allocated to the whole plane defined by the square (bounding rectangle) surrounding the outline islands: the land within islands had allocated a demand "1" (demand present), and the rest of the area "0" (no demand).

(80%) of transmitters are, from an engineering point of view, highly desirable design characteristics for wireless communication systems.

Concluding Remarks

We can sum up what we said so far about spatial evolutionary modeling as follows: when we want to model a spatial phenomenon using an evolutionary framework we follow three steps:

1. We create a model of a phenomenon—we call it a spatial model.
2. We convert this model into an evolutionary model with evolutionary framework—we create a spatial evolutionary model.
3. We construct the set of operators that follow the general principles of computational evolution but are specifically designed to operate on our model—these operators combined together into one algorithm, together with the underlying evolutionary model, are, what we call, a spatial evolutionary algorithm.

So far, our presentation of a spatial evolutionary algorithm has been rather informal. We intentionally omitted several aspects of the algorithm and the underlying model (genome structure, structure of operators, and a formal framework), providing only a cursory explanation of the operational details. The following sections provide technical details on all the aspects of the algorithm that have not been discussed in this introduction.

Spatial Evolutionary Model—More Details

In this section we will look more closely at the spatial evolutionary model and algorithm. We will discuss the design of chromosomes and the details of evolutionary operators. We will also give the complete pseudocode of the algorithm. At the end of this chapter we will provide the reader with an example that illustrates how the actual spatial evolutionary operators are implemented and how they function. The example is simple and can be validated with a handheld calculator.

As we have already pointed out, there are three phases in the construction of the evolutionary algorithm. In Phase 1 we create a model of spatial system. In Phase 2, we create a spatial evolutionary model. In Phase 3, we define evolutionary operators.

Phase 1 is rather straightforward and was discussed in sufficient detail in the previous section. In Phase 2, we create a spatial evolutionary model of the spatial model created in Phase 1. We have already stated that in our model, transmitters (and their service areas) are represented as organisms. A wireless communication system is represented as a population of organisms—each organism is a transmitter. Each organism has a genetic structure representing properties of a corresponding transmitter. A collection of populations is represented as a hyperpopulation. This hierarchical structure of the spatial evolutionary model (a hyperpopulation of populations composed of populations of organisms composed in turn of organisms with genomes) is illustrated in figure 2-23a. At the lowest level of the hierarchy are chromosomes that encode the organism-transmitter characteristics, such as *position, radius*, and *capacity to carry traffic.*

Coding of transmitter features is *natural.* That is, parameters of transmitters are *not* represented in a binary-coded format but in their actual values. On the next level are organisms and above them are populations of organisms. At the highest level—the root of the hierarchy—is a hyperpopulation that represents a collection of all populations at any given time. The spatial model that corresponds to this evolutionary model is given in figure 2-23b.

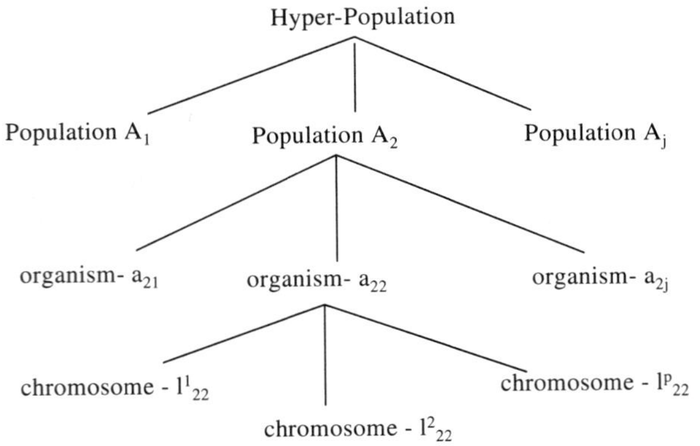

Figure 2-23a Structure of the genome of the evolutionary model of the wireless communication system.

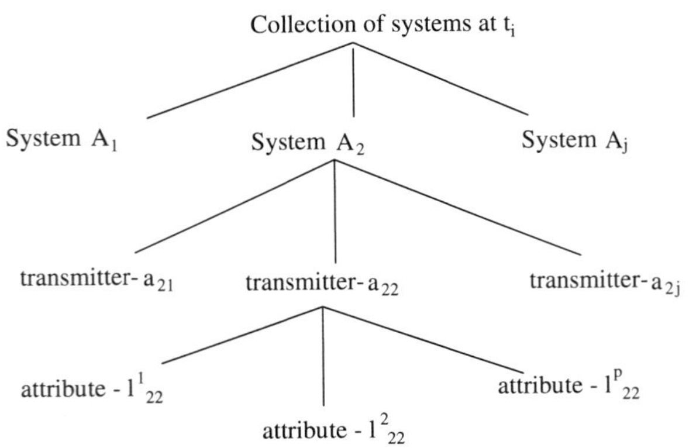

Figure 2-23b Representation of entities in the model of the wireless communication system.

The term "hyperpopulation" needs some explanation. We have stated that, in evolutionary language, a population is a collection of individuals that coexist in the same generation and undergo selection and reproduction. In our spatial evolutionary model, in each generation, there is a collection of populations of individuals. This *collection of populations* is logically equivalent to the *population* in a nonspatial evolutionary model. We use the term "hyper-population" to differentiate this collection of populations of individuals from the population of individuals in nonspatial evolutionary models. The differences are:

- A population in a nonspatial evolutionary model is a collection of individuals. These individuals store genetic material. They are under evolutionary pressure and evolve from generation to generation.
- A hyperpopulation in a spatial evolutionary model is a collection of populations. Each population in this collection is composed of several individuals. These individuals contain chromosomes. The genetic material of a population is a "sum," a total, of the genetic material of individuals that are in this population. The populations in a hyperpopulation are under evolutionary pressure and they evolve from generation to generation.

As we will see, the morphological differences between populations in spatial and nonspatial models go deeper than just the construct of the genome. These differences affect the way in which the operators are implemented. For example, in a nonspatial evolutionary model selection takes place between individuals. In a spatial evolutionary model, selection occurs between populations and between individuals. This duality of function is also true for mutation, cross-over, fitness, and mating.

Let us return for a moment to the design of the genome. In a more formal way, the structure of the genome of the spatial evolutionary model can be presented as follows (we use here a notation introduced in part I). The basic component of the spatial evolutionary model—the *organism*—is symbolized by a vector $\vec{\boldsymbol{a}}_{ij} = \{\boldsymbol{l}^1_{ij}, \ldots, \boldsymbol{l}^p_{ij}\}$, where the subscripts represent an organism i and population j, l^k_{ij} is the kth attribute of the ith organism in the jth population, and p is the number of attributes per organism. Thus, population $\boldsymbol{A}^t_j$ at time $\boldsymbol{t}$, may be represented as a vector $\{\boldsymbol{a}_{1j}, \ldots, \boldsymbol{a}_{nj}\}$ of dimension $\boldsymbol{n}$ (where $\boldsymbol{n}$ is the number of organisms in the population), or alternatively, as a matrix:

$$A^t_j = \left\{ \begin{matrix} l^1_{1j} & \ldots & l^p_{1j} \\ \ldots & \ldots & \ldots \\ l^1_{nj} & \ldots & l^p_{nj} \end{matrix} \right\}$$

of dimensions $\boldsymbol{p} \times \boldsymbol{n}$ [the number of attributes (chromosomes) in an organism times the number of organisms in the population] and the population of populations (hyperpopulation) is a matrix of dimension 3 of $\boldsymbol{p} \times \boldsymbol{n} \times \boldsymbol{m}$ where $\boldsymbol{m}$ is a number of populations. A complete description of the formal notation for the spatial evolutionary algorithm is presented at the end of this chapter.

We have not explained yet how spatial information is represented in a spatial evolutionary algorithm. In our model of a spatial evolutionary algorithm, spatial information about the distribution of demand, land use, and so forth, is represented as a set of data layers. The use of data layers, where each layer represents some particular aspect of the space (demand, land use, etc.) is a typical form of representation of spatial information in GISs. An example of such representation is demonstrated in figure 2-24. The map to the left of figure 2-24 symbolizes spatial information about a particular area. This collection may include information about roads, land use zones, the location of bodies of water, or demand for wireless services. Each one of these particular types of information is then abstracted as a separate "layer" of information (hence the "layered" model). The abstraction of spatial information layers is symbolically presented in figure 2-24 in the form of a layered stack of data.

Spatial Evolutionary Algorithm—Implementation Details

We can now proceed to Phase 3, the last step of the modeling process, in which we define evolutionary operators and consequently the spatial evolutionary algorithm (SEA). Our spatial evolutionary algorithm has been designed with seven operators: initialization, selection, fitness, mating and cross-over, mutation, learning, and objective function. The hierarchical architecture of a spatial evolutionary

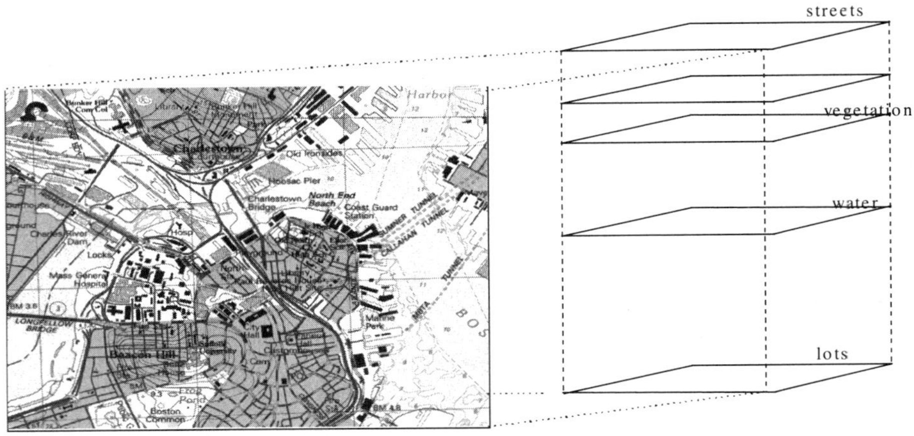

Figure 2-24 Symbolic representation of the "layered spatial data model" used in the spatial evolutionary model of a covering problem.

model imposes on each operator (with the exception of the objective function that acts only on the level of population) two levels of implementation: population and organism level.

An initialization operator generates an initial hyperpopulation. This initial hyper-population is a candidate, a starting solution of the model. From that initial solution, the evolutionary algorithm starts its search. Reflecting the hierarchical design of the evolutionary model of the wireless communication system, initialization is a hierarchical process involving generation of a hyperpopulation, populations, and organisms. A hyperpopulation is created by generating a predefined number of populations. A population is created by generating a predefined number of organisms. Organisms are generated by encoding their features in separate chromosomes. Each organism has at least two chromosomes representing its location in a 2-D geographic space.

Fitness operators measure the quality of organisms and populations with respect to the model objectives. Fitness provides a grouping (sequencing) score, which defines the standing of an organism relative to other organisms (organism fitness) or a population relative to other populations (population fitness). Fitness operators in the model of the wireless communication system are implemented as focal and zonal operators of map algebra (Tomlin, 1985, 1990).

Map algebra defines five classes of operators. They can be presented in a generic form as:

$$\langle \textit{output-layer} \rangle \langle \textit{opc} \rangle^{\langle \textit{cardinality} \rangle}_{\langle \textit{type} \rangle} \langle \textit{input}, \ldots, \textit{input} \rangle$$

where $\langle \textit{output-layer} \rangle$ is a map layer generated by the operation, $\langle \textit{opc} \rangle$ is a class of the map algebra operator, $\langle \textit{type} \rangle$ is a type of the operator, $\langle \textit{cardinality} \rangle$ is a number of layers the operator acts on, $\langle \textit{input}, \ldots, \textit{input} \rangle$ are layers the operator acts on. Only focal and zonal operators are of interest in our model, as focal operators express fitness of an organism, and zonal operators express fitness of a population. Further details were given above on spatial data structures.

A focal operator is defined as a metric over the subspace (area, volume) centered over a focal point. A metric is any operation (sum, average, median, etc.) on elements of space within the subspace of an operator. The subspace is either a regular geometric shape (circle, square, hexagon) or an arbitrary polygon. Parameters of a focal operator are: subspace type, metric type, and layer set upon which the operator is acting. Using the proposed notation, a focal operator can be represented as:

$$\langle o \rangle F \langle s|m \rangle \langle i \rangle$$

where $\langle o \rangle$ is an output layer, $\langle s| \rangle$ is a type of subspace, $\langle |m \rangle$ is a type of metric, and $\langle i \rangle$ is an input layer. $\langle i \rangle$ is any map layer compatible with the operand, $\langle s \rangle$ is either a regular shape or a polygon defined by its vertices, and $\langle m \rangle$ is a metric as defined above. The cardinality of a focal operator is "1" and it specifies that a focal operator uses one map layer as an input.

A zonal operator is a metric of the subspace, called a zone. A zone can be an area defined by a watershed, an area defined by a range of precipitation, an area defined by a zip code, or an area of a county. The zone may be one contiguous area or it may be several disconnected areas. A metric of a zonal operator can be any algebraic function (average, sum, median, minimum, maximum, etc.) permissible on attributes associated with locations within the zone. Parameters of the zonal operator are: a subspace, metric type, and layer set upon which the operator is acting. The zonal operator is not centered at any particular point as a focal operator is. A zonal operator is represented as:

$$\langle o \rangle Z \langle l|m \rangle \langle I \rangle$$

where $\langle o \rangle$ is an output layer $\langle l| \rangle$ is a map layer defining a zone, $\langle |m \rangle$ is a type of metric, and $\langle input \rangle$ as for a focal operator. The cardinality of zonal operator may be greater than 1.

The fitness of an organism determined by a focal operator is a function of the spatial environment, location of the organism in this environment as well as its location with respect to other organisms. The focal operator for the evolutionary model of a wireless communication system is defined as a sum of a demand assigned to points that belong to the service area of the transmitter.

The fitness of a population assesses the configuration of all the organisms constituting the population with respect to the underlying spatial information and the objectives of the evolution. In other words, population fitness is a measure of the collective—not individual—quality of all of the organisms constituting the population. The union of subspaces occupied by objects belonging to the population creates the zone of a zonal operator. The metric of a zonal operator defines the algorithm determining the fitness of a population. More formally, a zonal operator is defined as the sum of a demand assigned to points belonging to a zone $\langle l \rangle$—a sum of service areas (obj_i) of all the transmitters in one system (or a population using an evolutionary metaphor):

$$\langle l \rangle = \bigcup_{\substack{i=1 \\ i \in P}}^{m} obj_i$$

A selection operator selects the set of populations (and organisms) for reproduction (which is a combination of mating and cross-over operators). The spatial evolutionary model has two levels of selection: the level of populations and the level of organisms.[14] Two levels of the selection process reflect two levels of evolutionary units (i.e., populations and organisms).

Spatial selection of populations is a process of determining which populations from the hyperpopulation will be used for mating. Selection on this level is done using the population fitness. The actual mechanism of population selection is implementation dependent. On the level of organisms, selection is the process of determining which organisms will be used for mating. Spatial selection on this level is implicit. That is, the subset of organisms selected for mating is the subset that is included in populations selected for mating.

A selection process should produce populations with better fitnesses than these of populations in the previous generation. In the evolutionary model of a wireless communication system the population selection operator is implemented as Tournament 1-1 (Mitchell, 1996). In Tournament 1-1, two populations are randomly pooled from a hyperpopulation and, out of these two, the population with the greater fitness is selected for a new population. The spatial evolutionary algorithm selection mechanism for organisms is implicit, that is, organisms selected are those contained in selected populations.

A mating operator selects evolutionary units (or units of selection) for crossover (reproduction). Mating is a two-level process reflecting the two levels of evolutionary units. At the level of population, mating is a process of pooling of populations from a set of selected populations. The actual mechanism of pooling is implementation dependent. At the level of organisms, mating occurs between organisms of two populations selected for mating. Out of these two, one population is selected, at random. For every organism in this population a mate in the other population is selected, based on the "closeness" criteria. The definition of closeness is implementation dependent. This process is implemented as follows:

for $i : 0 \rightarrow n$

select a_{i1}

select $a_{kl} | D(a_{i1}, a_{k2}) = min$

cross-over

The above algorithm can be described as follows: for every organism i in population 1 (a_{i1}) select an organism in population 2 (a_{k2}) that is closest to it by some measure D. Our spatial evolutionary model uses the Euclidean distance metric as a measure of closeness. Other metrics can be designed as well.

A cross-over operator recombines the genetic material between organisms of mating populations in order to produce different (and hopefully better) organisms. In our evolutionary model, cross-over is a two-level process reflecting the two levels of evolutionary units. At the level of populations, cross-over between populations is implicit. The cross-over of populations takes, as its input, *two* populations and generates *one* offspring population. The cross-over process between organisms is a process in which a chromosome responsible for

Feature 1 in Organism A from one population is combined with a chromosome responsible for Feature 1 in Organism B from another population in order to produce a chromosome responsible for Feature 1 in an offspring organism in the offspring population. The cross-over mechanism is implementation dependent and chromosome specific, to preserve its closure.

In the spatial evolutionary model of a wireless communication system a *cross-over* operator for organisms is implemented as a fixed-point spatial cross-over and is defined by the following formula:

$$l_i^k = w^1 l_1^k + w^2 l_2^k$$

where weights w^1 and w^2 are equal to 0.5. The l_i^k are attributes representing genes of organisms used by the cross-over operator. This formula states that genes that are crossed-over represent the same attributes and the resulting gene also represents the same attribute. That is, a gene representing coordinate X of the location of organism 1 is crossed over with a gene representing coordinate X of the location of organism 2, and the resulting gene represents coordinate X of the location of organism i.

A mutation operator is implemented as two operators: "Big M" mutation and "small m" mutation. Big M mutation affects the whole population. It is implemented as a random change in the genetic material of all organisms in a randomly selected population. "Small m" mutation is a random change in the genetic material of the randomly selected organism in a randomly selected population.

Mutation in the spatial evolutionary algorithm is different from the mutation in nonspatial evolutionary algorithms that use binary coding. It acts on the value represented by a gene—in its problem space—and it affects the whole chromosome. In contrast, mutation in other implementations of evolutionary algorithms (particularly in those which are binary coded) acts on a particular location, equivalent to a bit, of the chromosome. One may say that mutation in binary-coded representations is chromosome driven rather than problem driven, as in evolutionary models using a natural representation.

Mutation for location chromosomes defined over a metric space is implemented as:

$$\begin{aligned}
&\text{if } u_1(0,1) > x_{old}/(x_{max} - x_{min})\\
&x_{new} \leftarrow x_{old} + u_2(0,1) * (x_{max} - x_{old})\\
&\text{else}\\
&x_{new} \leftarrow x_{old} - u_2(0,1) * (x_{old} - x_{min})
\end{aligned}$$

where $u(0, 1)$ is a uniform random variable between 0 and 1, x_{old} is a chromosome undergoing mutation, x_{new} is a mutated chromosome, x_{max} is the maximal bound on a value of chromosome, and x_{min} is the minimal bound on the value of chromosome. This algorithm preserves the closure of the mutation and depends on the current state of the chromosome.

A learning operator[15] is implemented as a local, deterministic hill-climbing algorithm, which is represented as follows:

$$
\begin{aligned}
&\text{for } i : 0 \rightarrow n\{\\
&\quad \text{do}\{\\
&\quad\quad \text{do}\{\\
&\quad\quad\quad move(x_i, y_i, d, s)\\
&\quad\quad\quad obj(I_l)\\
&\quad\quad \}while(\Delta obj(I_l) > 0)\\
&\quad\quad change\_direction(d) \text{ or}\\
&\quad\quad change\_step(s)\\
&\quad \}while(s > min(s))\\
&\}
\end{aligned}
$$

where *move*() expresses a search from a point $I_i(x, y)$, with a step s in a direction d, $obj(I_i)$ is an objective function value for an object I_i, $min(s)$ is a minimum step, $\Delta obj(I_l)$ is a change in a objective function value, *change_direction*(*d*) is a change in the direction of a search, *change_step*(*s*) is a change in the step of search, and n is a maximum number of iterations allowed. A learning operator acts between generations on selected organisms in selected populations. A number of populations selected for learning, a number of organisms that learn, and a number of "learning cycles" are parameters of a learning operator. Populations and organisms used for learning are selected randomly from the populations in a given hyperpopulation.

An objective function in a spatial evolutionary algorithm is defined only for populations (as only the population fitness maps into the problem space) and is equivalent to the fitness of the population. It expresses the fitness of the evolutionary units—populations—with respect to the problem objectives. In the evolutionary model of a wireless communication system a fitness function expresses the total traffic demand covered by the transmitters.

Each of the operators in a spatial evolutionary algorithm has one or more parameters determining its activation levels or other operational characteristics. The description of those parameters is omitted as it is not essential for the understanding of an algorithm.

Spatial Evolutionary Algorithm—a Code

The pseudocode of the spatial evolutionary algorithm (SEA) implemented to model the wireless communication system is presented in table 2-4. The SEA is composed of two segments: an initialization segment and a main loop segment (*while .. end while*). In the initialization segment there are two functional blocks: initialization of populations and fitness calculation. Evolution is carried out in the

Table 2-4 A pseudo-code for the spatial evolutionary algorithm

```
for i:0 → m
  for j:0 → n
    initialize m populations of n objects each()
    // GET FITNESS SCORES
    get fitness for n objects for m populations()
    get fitness for m populations()
  WHILE (STOP == NO)
    // SELECT POPULATIONS FOR MATING
    select m' populations from m()
    // MATE POPULATIONS
    for i: 0 → m'{
                      get population I
                      select mating population k()
                      mate objects of population i with objects
                                    of population k ()
     }
    // MUTATION (rnd is random number → (0, 1)
    if rnd < Prob of Big M
                      select population()
                      mutate()
    if rnd < prob of small m
                      select population()
                      select object()
                      mutate()
    // LEARNING
    if rnd < prob of learning
                      select number of learning populations l()
                      for i:0 →l
                               ford: 0 → ll // learning levels
                                                    learn poulation i
    // GET NEW FITNESS
    get fitness for n objects for m populations()
    get fitness for m populations()
    if stopping criteria reached()
                      STOP = YES
ENDWHILE
```

main loop. This loop contains selection, mating, cross-over, fitness calculation, small mutation, big mutation, and learning operators. Selection, mating, cross-over, and fitness calculations are activated at each evolution cycle. Small mutation, big mutation, and learning are activated if the pooled random number (*rnd*) is less than the activation probability for a given operator. The loop (*while ... end while*) is terminated when the stopping criterion is met.

Spatial Evolutionary Algorithm—an Example

It will be easier to understand the discussed concepts of the genome structure and operators if we demonstrate how they can be translated into a model of a real-life problem. For this purpose we have selected a very simple model that is easy enough to be designed on the back-of-an-envelope, but has most elements that more complex models have. The example we use is a model of a wireless communication system with three transmitters.

A simple, three-transmitter wireless communication system in our spatial evolutionary model would be represented as a population with three individuals as follows:

$$(10.0, 15.0, R), (20.0, 30.0, R), (2.0, 3.0, R)$$

A single individual—a transmitter—has three chromosomes: one for coordinate X, one for coordinate Y, and one for radius R (which is constant). The first individual (10.0, 15.0, R) is a transmitter located at the location with coordinates $x = 10.0$ and $y = 15.0$, with a service area radius of R. The second individual (20.0, 30.0, R) is located at $x = 20.0$, and $y = 30.0$, with a service area radius of R. The third individual is (2.0, 3.0, R).

Two populations of three transmitters each, are represented in our model as follows:

$$[\{(10.0, 15.0, R), (20.0, 30.0, R), (2.0, 3.0, R)\};$$
$$\{(7.0, 5.0, R), (15.0, 2.0, R), (20.0, 14.0, R)\}]$$

where "{ }" brackets enclose a population, "( )" brackets enclose a single transmitter-individual, and "[ ]" brackets enclose a hyperpopulation. We assume these are two populations selected for reproduction, that is, mating and cross-over. The matrix below presents the distances between individual of these two populations, calculated using the Euclidean distance metric: the distance metric is needed to select individuals for cross-over.

	(7, 5, R)[2, 1]	(15, 2, R)[2, 2]	(20, 14, R)[2, 3]
(10, 15, R) [1, 1]	10.44	13.92	**10.04**
(20, 30, R) [1, 2]	18.17	28.44	**16.00**
(2, 3, R) [1, 3]	**5.38**	13.03	21.09

The symbols in the square brackets next to individuals denote individual's population and organism index. For example the index [1, 2] next to the individual (20, 30, R) means that this individual has an index "2" within a population and that he is from a population "1." Given this distance matrix using the mating operator the following pairs of individuals will be selected for cross-over: [1, 1] with be crossed over with [2, 3], [1, 2] with [2, 3], and [1, 3] with [2, 1]. In the distance matrix, distances for selected pairs of individuals are in bold.

The cross-over operator is an arithmetic average and, on these individuals, will produce the following offspring individuals:

$$(10.0, 15.0, R)(\times)(15.0, 2.0, R) = (12.5, 8.5, R)$$

$$(20.0, 30.0, R)(\times)(15.0, 2.0, R) = (17.5, 16.0, R)$$

$$(2.0, 3.0, R)(\times)(20.0, 14.0, R) = (11.0, 8.5, R)$$

Table 2-5 A comparison of three levels of representation of evolutionary models: conceptual, formal, and implementation

Concept	Formal Representation	Implementation
Chromosome	L_{ij}^k	10.0
Individual	$\vec{a}_{ij} = \{I_{ij}^1, \ldots, I_{ij}^p\}$	$(10.0, 15.0, R)$
Population	A_j^t	$(10.0, 15.0, R), (20.0, 30.0, R), (2.0, 3.0, R)$
Hyperpopulation	A matrix of dimension 3 of $p \times n \times m$ where m is the number of populations	$(10.0, 15.0, R), (20.0, 30.0, R), (2.0, 3.0, R)$ $(7.0, 5.0, R), (15.0, 2.0, R), (20.0, 14.0, R)$

$\{(12.5, 8.5, R)(17.5, 16.0, R)(11.0, 8.5, R)\}$ is a new population after cross-over. The same process of mating and cross-over would be repeated for another pair of populations, until a complete hyperpopulation has been created.

It may be difficult, at first, to see how the particular evolutionary model, such as the model given above, relates to the abstract representation of a model introduced earlier. Let us, then, go over an example that would elucidate this relation. An individual (10.0, 15.0, R) is equivalent to $\vec{a}_{ij} = \{I_{ij}^1, ..., I_{ij}^p\}$. For this individual an index p is equal to 3. A population $\{(12.5, 8.5, R)(17.5, 16.0, R)(11.0, 8.5, R)\}$ is equivalent to $\{a_{1j}, \ldots, a_{nj}\}$. The matrix form of a population

$$A_j^t = \begin{Bmatrix} l_{1j}^1 & \cdots & l_{1j}^p \\ \cdots & \cdots & \cdots \\ l_{nj}^1 & \cdots & l_{nj}^p \end{Bmatrix}$$

represents the population:

$$\begin{matrix} 12.5, 8.5, R \\ 17.5, 16.0, R \\ 11.0, 8.5, R \end{matrix}$$

A hyperpopulation

$$A_j^t = \begin{Bmatrix} l_{1j}^1 & \cdots & l_{1j}^p \\ \cdots & \cdots & \cdots \\ l_{nj}^1 & \cdots & l_{nj}^p \end{Bmatrix}, \quad A_j^t = \begin{Bmatrix} l_{1j}^1 & \cdots & l_{1j}^p \\ \cdots & \cdots & \cdots \\ l_{nj}^1 & \cdots & l_{nj}^p \end{Bmatrix} \cdots$$

represents

$$\begin{matrix} 10.0, 15.0R & 7.0, 5.0, R \\ 20.0, 30.0, R & 15.0, 2.0, R \\ 2.0, 3.0, R & 20.0, 14.0, R \end{matrix}$$

Formal and model-specific forms of representation of a spatial evolutionary model (as well as a conceptual representation) are compared side by side in table 2-5.

Spatial Evolutionary Algorithm—Properties, Behavior, Parameters

Properties

The performance of the SEA has been tested against two other algorithms—the random placement algorithm (RPA) and Tornqvist's algorithm (TA). The RPA is similar to the initialization phase of a spatial evolutionary algorithm. For a given spatial data set and model, a collection of facilities is generated at random locations. The number of generated facilities is equal to the number of organisms in one population of an equivalent spatial evolutionary model. The same restrictions on the locations of facilities apply as in the initialization phase of a spatial evolutionary model (i.e., facilities have to be within the bounding box of a data set). For this collection of facilities, the objective function (the same as for the equivalent spatial evolutionary model) is calculated. This process is repeated a number of times and the best result is retained as a solution of an algorithm run.

The TA is a deterministic algorithm belonging to a family of local hill-climbing search methods. These methods involve a series of moves over the search space, with each move attempting to improve the objective function. The TA was first formulated 25 years ago (Tornqvist et al., 1971) but it is still regarded as the best existing heuristic for planar covering problems (Goodchild, 1984).

In all tests performed in this study, the SEA demonstrated superior results to both the RPA and TA. In comparisons of individual tests, the SEA results were up to 25 percent higher than the *best* results acquired by the TA and up to 30 percent higher than the *best* results for the RPA. These differences were even wider when *average* results were compared.

Spatial Evolutionary Algorithm–Behavior

In this section we elucidate a mechanism of evolutionary search and roles of evolutionary operators in the spatial evolutionary algorithm. For this purpose we perform an experiment with a simple covering problem. The problem requires us to cover four facilities with the maximum amount of demand over a 3×3 squares checker-board data set; figure 2-25 represents the data set. Only four squares out of nine in the checker-board have assigned demand. Thus these four facilities must be located at these squares. In figure 2-25 the squares with no demand are gray and squares with demand assigned are white.

The evolutionary model used in the experiment is the same as the one elaborated in the preceding sections, that is, the genome of the evolutionary model is hierarchical; each individual symbolizing a facility has three chromosomes: one for an x location, one for an y location, and one for a constant radius R. The radius R is equal to one-half of the side of a demand square. Individuals are grouped into populations: one population represents one configuration of facilities. Populations are grouped into a hyperpopulation. All the operators in this evolutionary model are also the same as discussed before.

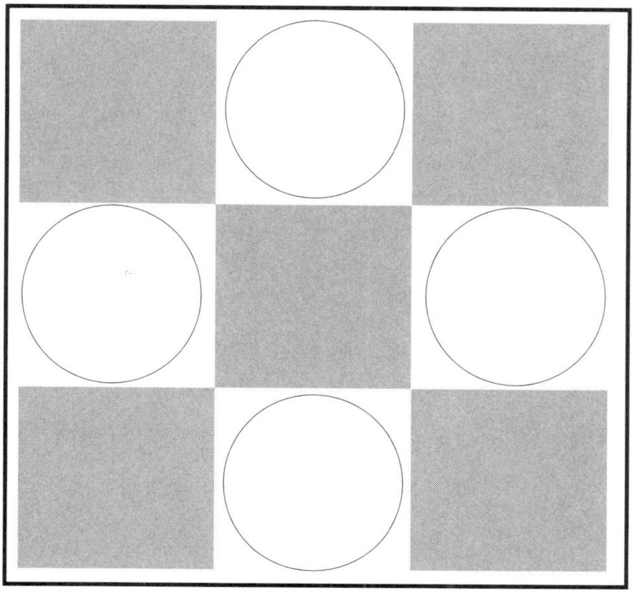

Shaded squares represent areas without allocated demand
Four circles represent the optimal location of four "facilities"

Figure 2-25 Spatial data set for a SEA performance study.

The problem modeled in our experiment is, in fact, a very simplified version of a transmitter location problem in a model of the wireless communication system. Because the problem is so simple, we can observe the progress of evolutionary search from a generation to a generation. Moreover, since we a priori know the solution to the problem, we can also evaluate the quality of the solution provided by our model.

During the experiment, at each generation we recorded positions of individuals in every population, and four evolutionary statistics: minimum, maximum, average fitness scores, and the difference between the maximum and the minimum fitness scores. The minimum and maximum population fitness scores reflect the fitness scores of the best and the worst population at each generation. The difference between these scores reflects the spread of the fitness scores in the hyper-population. This statistics would be expected to converge to almost 0.0 towards the end of evolution as the populations converge towards one optimum configuration. For the same reason the average fitness score statistics would be expected to converge towards the maximum fitness score statistics.

The model was run until no further improvement in the model objective function could have been achieved. The results of this experiment are summarized in figures 2-26a and b and 2-27. Figures 2-26a and b present the locations of all four facilities at each generation (1–12). We can see that the evolutionary search starts at random locations probing the whole demand surface. With each next generation, the search progressively focuses upon increasingly smaller areas until it converges to locations at the centers of four squares with assigned demand.

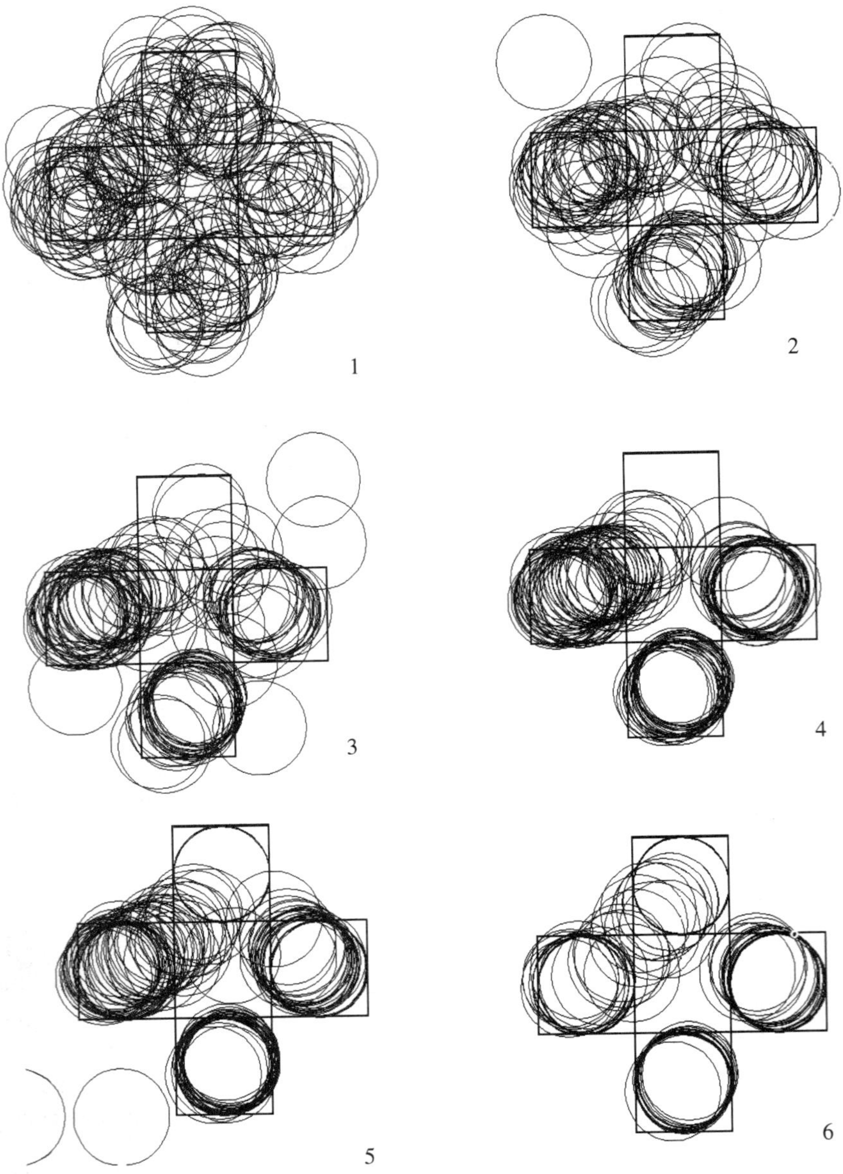

The number in the lower right corner shows the cycle number. Outlines of circles represent all 160 objects in 40 populations.

Figure 2-26a Location of objects in the test in cycles 1 to 6.

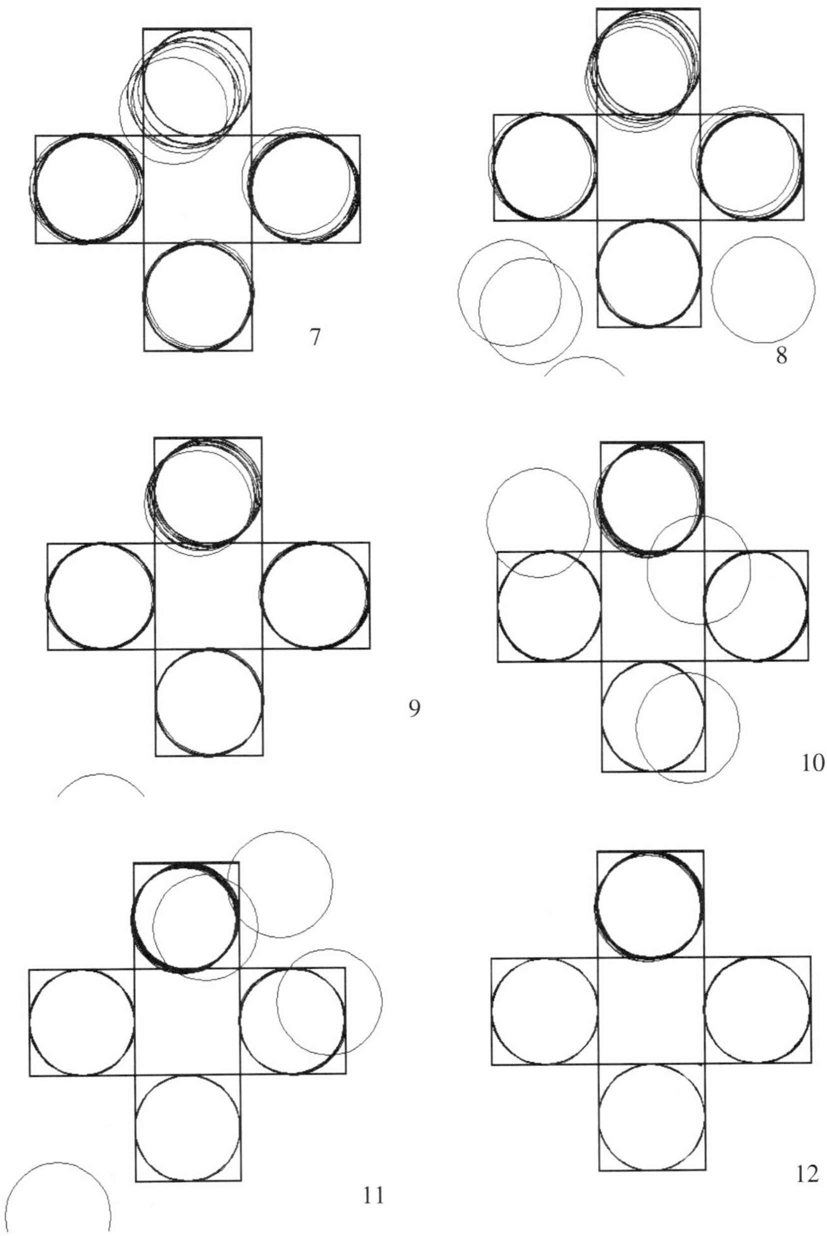

The number in the lower right corner shows the cycle number. Outlines of circles represent all 160 objects in 40 populations. (Cont.)

Figure 2-26b Location of objects in the test in cycles 7 to 12.

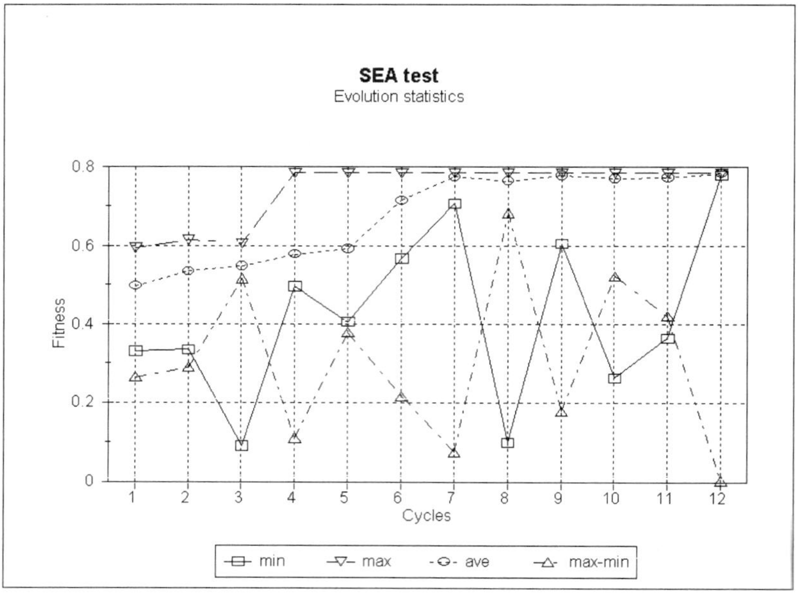

A curve of minimum fitness shows randomness in values and no correlation to the curve of maximum and average fitness. On those curves two plateaus are visible.

Figure 2-27 Evolution statistics (maximum, minimum, average, and maximum–minimum) for the test.

Figure 2-27 presents four evolutionary statistics at each generation. The analysis of changes of these evolutionary statistics during evolution allows for several interesting observations about the behavior of the spatial evolutionary algorithm:

- The evolution for our model lasted for 12 generations. The final solution to our model produced by evolution agreed with what we expected to obtain—the four facilities have been located at the centers of four squares with assigned demand.
- The average fitness score (the second line from the top in figure 2-27) was growing monotonously during the whole evolution with an exception of intervals between the third and fifth generation and the ninth and twelfth generation. During these generations the average fitness score did not change at all.
- The minimum fitness scores (the second line from the bottom in figure 2-27) varied considerably from a generation to a generation. These variations are also reflected by the difference between the maximum and minimum population fitness (the bottom line in figure 2-27). But they are not reflected by the maximum fitness score statistics.
- The maximum population fitness scores (the top line in figure 2-27) were monotonously increasing to the fifth generation with an exception of a plateau from the second to the third generation. From the fifth generation the maximum population fitness score did not change.

The plateaus on the graph of the average fitness scores suggest that the evolutionary search converges at some generations at a local optimum. The increase in fitness *after* the plateaus suggests that an evolutionary search can break out of the local optima.[16] Further, it could be proposed that the occurrence of plateaus has been a result of the combined action of selection/cross-over and learning operators. The selection/cross-over operators are interpolators. They cannot sample areas outside the convex defined by the locations of existing populations. Thus, the population tends to very quickly converge to a local optimum, even if this optimum is local, not global. A learning operator is an extrapolator. It searches outside of the domain defined by populations. Once it finds better regions, it extends the domain defined by populations and populations may move to the new optimum using selection and cross-over operators.

The variations in the minimum fitness scores could be explained by the action of the mutation operator. Every time the mutation operator is activated it introduces the new poorly scoring population. However, as the average and the maximum fitness statistics show, this disruptive action of the mutation operator does not affect the convergence of the populations. We could say that the mutation operator had no affect on the evolution. This observation was confirmed by further studies (Krzanowski, 1997).

Based upon the findings of this experiment, the following comprehensive description of spatial evolutionary search is proposed. Spatial evolutionary search starts at random locations distributed all over a 2-D plane during the initialization process. As generations go by, the search is narrowed to increasingly smaller areas of the plane until the search focuses on the optimum locations. The observed search process might be compared to focusing of lenses. There is more to this analogy than just a superficial similarity. The "focusing of lenses" analogy tells us that the evolutionary search looks the problem space as a whole and then "focuses on" selected areas of this space that include the optimum locations. We should contract this search process with a search employed by hill-climbing algorithms—algorithms most often used to solve spatial search problems—that probe the problem space one point at the time.

Parameters

Several experiments have been conducted to elucidate the role the SEA parameters and operators (Krzanowski, 1997). As these experiments have been quite elaborate only the most important results are given here. Table 2-6 gives the range and types of parameters and settings tested for the SEA. Experiments have been conducted on the checker-board data set reported in a previous section.

Table 2-7 presents the recommended settings for the parameters of the operators and the selected types of operators. Again the full details of experiments are given in Krzanowski (1997). Two observations from Table 2-7 are of particular interest as they contradict to some extent what one would expect knowing non-spatial evolutionary models: first, that mutation (both big M and small m) within the range of tested parameters was demonstrated to have no effect on the evolution, and second that the learning algorithm was beneficial to evolution only if it was activated once or at most twice during the run of the model.

Mutation has been always considered as an important diversification operator in the genetic search. There are some evolutionary models (as the reader remembers

Table 2-6 Range of parameters of the SEA tested in the performance study

Parameter	Levels
Population size	10, 20, 40, 80
Learning cycles[a]	1, 2, 3, 4
Selection[b]	1, 2
Cross-over[a]	1, 2, 3, 4
Probability of big mutation	0.05, 0.1, 0.2, 0.3
Probability of small mutation	0.05, 0.1, 0.2, 0.3
Learning probability	0.01, 0.2, 0.3, 0.4
Learning rate	0.01, 0.2, 0.3, 0.4

[a]1: Fixed point; 2: random; 3: local fitness; 4: global fitness.
[b]1: Tournament 1-1; 2: roulette wheel with power scaling.

from the introduction) that use only mutation. However, in the modeling of spatial problems mutation seems to have no influence on the evolutionary search at all. Closer analysis of the model runs demonstrated (Krzanowski, 1997) that the mutation did not generate at any time any populations that would have improved fitness. To the contrary, mutated populations had always lower fitness than the populations before the mutation.

The effect of the learning algorithm on the evolutionary search is also a surprise. One would expect that the learning algorithm should benefit the evolutionary model more, if it is more often applied. As tests demonstrated, "more learning" translated into higher learning probability and higher learning rate does not translate into a better model. There is a level of "learning" at which the model converges to the best solution. Beyond or below this level, learning has no effect on the evolutionary search. Could we call this behavior a "learner's block"? Obviously these results hold only for the tested learning algorithm.

Table 2-7 Recommended parameter settings for the SEA

Parameter	Significance	Recommended Settings	Range of Values or Settings
Population size	Yes	40, 80	> 40
Learning cycles	No (if > 1)	1 or 2	≥ 1 No effect of more than two learning cycles
Selection method	No	Tournament 1-1	
Cross-over method	Yes	Fixed, or fitness based	All but random cross-over
Big mutation probability	No		
Small mutation probability	No		
Learning probability	Yes	0.2	≥ 0.2 No effect of higher tested probabilities
Learning rate	Yes	0.2	≥ 0.2 No effect of higher tested probabilities

Spatial Evolutionary Algorithm: Formal Framework

In the preceding discussion, the authors have only briefly touched upon formal aspects of the spatial evolutionary model. At this point, we will develop a comprehensively detailed, formal framework. The framework for spatial evolutionary algorithms defines concepts that are independent of the particular implementation of the algorithm and might be regarded as common elements of any spatial evolutionary model, regardless of the particular application.

The basic component of the SEA—the *organism*—is represented as a vector $\vec{a}_{ij} = \{\boldsymbol{l}^1_{ij}, \ldots, \boldsymbol{l}^p_{ij}\}$, where l^k_{ij} is the kth attribute of the ith organism in the jth population, and p is the number of attributes per organism. Each organism a belongs to the space of organisms O. Each attribute l^k is defined over its own attribute space L^k, where L may be any space except for attributes 1 and 2 (and 3 in a 3-D space), which are arbitrarily defined over the Euclidean space: these attributes define a location of an organism. A^t_j is a j population of organisms at time t. Γ is a space of all populations A_j. Γ^t is a subset of Γ at time (generation) t. A population A^t_j may be represented as vector of $\boldsymbol{a}_{ij}$, or a matrix of a_{ij} expanded to its components, each row being an organism. Thus, population $\boldsymbol{A}^t_j$ may be represented as $\{\boldsymbol{a}_{1j}, \ldots, \boldsymbol{a}_{nj}\}$ a vector of dimension $\boldsymbol{n}$, where $\boldsymbol{n}$ is the size of population with n being the number of organisms in a population. Alternatively, a population $\boldsymbol{A}^t_j$ can be represented as a matrix:

$$A^t_j = \begin{Bmatrix} l^1_{1j} & \ldots & l^p_{1j} \\ \ldots & \ldots & \ldots \\ l^1_{nj} & \ldots & l^p_{nj} \end{Bmatrix}$$

of dimensions $\boldsymbol{p} \times \boldsymbol{n}$ (the number of attributes – chromosomes in an organism $\times$ the number of organisms in a population). A space of populations at time $\boldsymbol{t}$, Γ^t is composed of $\{\boldsymbol{A}^t_1, \ldots, \boldsymbol{A}^t_m\}\boldsymbol{m}$ populations $\boldsymbol{A}$. The population of populations (hyperpopulation) is a matrix of dimension 3 of $\boldsymbol{p} \times \boldsymbol{n} \times \boldsymbol{m}$ where $\boldsymbol{m}$ is a number of populations and indices $\boldsymbol{p}$ and $\boldsymbol{n}$ are defined as before. We assume in this work that:

$$p = const. \bigwedge_{i,j,t} a_{ij} \in O$$

and

$$n = const. \bigwedge_{j,t} A^t_j \in \Gamma$$

which means that number of chromosomes in an organism and the number of organisms in populations is constant across all populations and generations.

The extension of this case, such as

$$p \neq const. \bigwedge_{i,j,t} a_{ij} \in O$$

and

$$n \neq const. \bigwedge_{j,t} A^t_j \in \Gamma$$

can be easily defined. The assumption of the constant population size and of the constant genetic make-up is implied in the most of the implementations of evolutionary algorithms.

Fitness

The fitness of organism a is defined as the function mapping $\phi{:}O \rightarrow \Re$ from a space of individuals into the real space $\Re$:

$$\phi{:}\phi(a_{ij}) \rightarrow \Re$$

One can write an organism fitness function as a multidimensional function defined over a set of spaces $\boldsymbol{L}^1, \ldots \boldsymbol{L}^p$, mapping them into $\Re$ as follows:

$$\phi{:}\phi\{l_{ij}^1, \ldots, l_{ij}^p | l_{ij}^1 \in L^1, \ldots, l_{ij}^p \in L^p\} \rightarrow \Re$$

Each space is the specific domain space of the specific feature of an organism.

The fitness of a population is defined as $\Phi{:}\Gamma \rightarrow \Re$ and is a mapping from a space of populations into a real space. Fitness function Φ determines the fitness of the population $\boldsymbol{A}_j^t$ and it is a function of $\boldsymbol{a}_{ij}$, that is, $\Phi(A_{ij}) = \Phi(a_{1j}, \ldots, a_{nj})$ and can be written as:

$$\Phi{:}\Phi(A_{ij}) \rightarrow \Re$$

The following relation between the fitness of organisms and the fitness of a population is also defined:

$$\sum_{i=1}^{n} \phi(a_{ij}^t) \neq \Phi(A_j^t)$$

This formula expresses the fact that the fitness of population is not equal to the sum of fitness of organisms—a fundamental condition of two level evolutionary units.

Selection

Formally, the population selection S_Θ is a function on the set Γ^t (at generation $\boldsymbol{t}$) of populations $\boldsymbol{A}$. It is defined as a mapping $S_\Theta : \Gamma^t \rightarrow \Gamma^{t*}$ of the initial population into the set of population of selected populations. $()_\Theta$ is a parameter of the selection operation.

Mating

Mating, on a population level, is a mapping $T_\Theta : \Gamma^{T*} \rightarrow \{A_j^t, A_k^t\}$, $j \neq k$. That is, the mating operator selects pairs of populations from the subset of selected populations such that the two selected populations are not the same. Mating on an organism level is a mapping

$$t_\Theta : \{A_i^t, A_k^t\} \rightarrow \{a_{ij}^t, a_{kl}^t | i \neq k \wedge F_\Theta(a_i^t, a_k^t) \leq D\}$$

for $a_{ij}^t \in A_i^t, j = 1, \ldots, n$. The function F_Θ is a function defining a measure of similarity or proximity between organisms and D is its limiting value. F_Θ is implementation (and problem) dependent.

Cross-Over

Formally, the cross-over on the populations level is defined as $\boldsymbol{R}_\Theta\colon\{\boldsymbol{A}_i^t, \boldsymbol{A}_k^t\} \rightarrow \boldsymbol{A}_l^{t*}$. On the organismic level the cross-over operator is defined as:

$$r_\Theta : \{A_{ij}^t, A_{kl}^t\} \rightarrow \{A_{hg}^{t*} | l_{h1}^{t*} = f_1(l_{i1}^t, l_{k1}^t), \ldots l_{hn}^{t*} = f_n(l_{in}^t, l_{kn}^t)\}$$

which expresses the fact that every gene has its specific cross-over operator $f_j(l_{i1}^t, l_{k1}^t)$.

Mutation

There are two types of mutation in the spatial evolutionary model: big M mutation and small m mutation. Formally, big M mutation is a random change to every organism and every gene of a population pooled for mutation. Big M mutation is defined as follows $M_\Theta : \Gamma^{t*} \rightarrow \Gamma^{t**}$, and small m mutation is defined as $m_\Theta : \Gamma^{t*(*)} \rightarrow \Gamma^{t***}$. In more detail, if a population of populations is:

$$\Gamma^t = A_1^t, \ldots, A_h^{t*}, \left\{ \begin{array}{ccc} 1_{1h}^1 & \ldots & 1_{1h}^p \\ \ldots & \ldots & \ldots \\ 1_{nh}^1 & \ldots & 1_{nh}^p \end{array} \right\}, \ldots, A_m^t$$

before big M mutation, after big M mutation it is:

$$\Gamma^{t**} = A_1^t, \ldots, A_h^{t*}, \left\{ \begin{array}{ccc} 1_{1h}^{1*} & \ldots & 1_{1h}^{p*} \\ \ldots & \ldots & \ldots \\ 1_{nh}^{1*} & \ldots & 1_{nh}^{p*} \end{array} \right\}, \ldots A_m^t$$

where A_h^* is a mutated population in which every gene of the mutated population is affected by the mutation.

Small m mutation is a random change to a genetic material of a selected organism of a selected population. More formally, if a population before the mutation is:

$$\Gamma^t = A_1^t, \ldots, A_{ih}^t, \left\{ \begin{array}{ccc} l_{im}^1 & \ldots & l_{im}^p \\ l_{ig}^1 & \ldots & l_{ig}^p \\ l_{im}^1 & \ldots & l_{im}^p \end{array} \right\}, \ldots, A_m^t$$

the same population after the small mutation is:

$$\Gamma^{t***} = A_1^t, \ldots, A_{ih}^{t***} \left\{ \begin{array}{ccc} \ldots & \ldots & \ldots \\ l_{ig}^{1***} & \ldots & l_{ig}^{p***} \\ \ldots & \ldots & \ldots \end{array} \right\}, \ldots, A_m^t$$

where A^{t***} is a mutated organism.

Learning

Formally, the learning process is defined as follows:

$$L_\Theta : A_h^{t()} \rightarrow A_h^{t****} \Rightarrow \{A_{hi}^{t****} | \phi(A_{hi}^{t****}) \geq \phi(A_{hi}^{t()}, i = 1, \ldots, m\}$$

This formulation does not define the particularities of the learning process but defines the condition that characterizes it. The particular implementation of learning in SEA is problem specific.

Objective Function

Objective function is defined only for populations (as population, alone, maps into the problem space) and expresses the fitness of the hyperpopulation with respect to the problem (represented by the evolutionary model) objective. An objective function is defined as

$$f : \Re^z \rightarrow \Re$$

where $\Re$ is a problem space. Objective function mapping the population fitness into the problem space is generally equivalent to the population fitness function Φ.

Termination Criteria

Termination criteria are denoted as $\iota :: \Gamma^t \rightarrow \{true, false\}$, which is *true* if the particular objectives of evolution are achieved by the population at a generation t, and *false* otherwise. Termination criteria are applied to population fitness only.

Spatial Evolutionary Algorithm—a Complete View

Using the notation and concepts defined in earlier sections, the complete SEA framework is represented as follows:

$t := 0$

initialize Γ^0

evaluate $\phi : \phi(a_{ij}) \rightarrow \Re$

$\Phi : \Phi(A_{ij}) \rightarrow \Re$

while($\iota :: \Gamma^t \neq true$){

select	$S_\Theta:\Gamma^t \to \Gamma^{t*}$
mate	$T_\Theta:\Gamma^{T*} \to \{A_j^t, A_k^t\},\ j \neq k$
	$t_\Theta:\{A_i^t, A_k^t\} \to \{A_{ij}^t\}$
reproduce	$R_\Theta:\{A_i^t, A_k^t\} \to A_l^{t*}$
mutate	$M_\Theta:\Gamma^{t(*)} \to \Gamma^{t(*)*}$
	$m_\Theta:\Gamma^{t(*)} \to \Gamma^{t(*)*}$
learn	$L_\Theta:A_h^{t(*)} \to A_h^{t(*)*}$
evaluate	$\phi:\phi(a_{ij}) \to \Re$
	$\Phi:\Phi(A_{ij}) \to \Re$
$t := t + 1$	

}

where O is a space of organisms and Γ is a space of populations of organisms. O^t and Γ^t are subsets of space of organisms and populations at time t. p is the number of chromosomes per organisms. m is the number of populations and n is the number of organisms per population. Φ is a population fitness operator. ϕ is an organism fitness operator. S_Θ is a selection operator. T_Θ is a population mating operator, t_Θ is an organism mating operator. m_Θ is a small mutation operator. M_Θ is a big mutation operator. L_Θ is a learning operator. ι is a termination criterion. In all cases, Θ means a parameter set of a specific operator.

Concluding Remarks About Spatial Evolutionary Algorithms

Spatial evolutionary algorithms present a framework for the modeling of spatial problems using evolutionary processing methods. The SEA framework consists of concepts and procedures that allow the creation of spatial evolutionary models of a realistic spatial system. Further, it encompasses the algorithms that allow one to manipulate the elements of the models.

Spatial evolutionary algorithms can be also defined as a mechanism for expressing models of spatial problems in an evolutionary framework, with elements of this framework that could be both manipulated and interpreted as spatial objects and evolving units.

The SEA offers the advantages of spatial information systems and evolutionary methods combined. Viewed from the vantage point of evolutionary models, the SEA offers an integrated, knowledge-rich environment, involving both spatial data and spatial models. In such a hybridized environment, evolutionary models are considered to perform at their best. From the perspective of spatial information systems, the SEA offers a new modeling method that is fully integrated, both at a low level of problem representation, as well as at a conceptual level with spatial models.

As experiments demonstrated, the SEA permits modeling of complex problems that, heretofore, might have been considered beyond the scope of traditional modeling methods either because of their complexity or because of the inability of traditional models to represent complex information on which these models depend.

It is a truism that "good research generates more questions that it answers." While this research project has articulated a promising new method for modeling of spatial problems, there are, indeed, many questions for which answers are not yet available. The following list presents various issues related to spatial evolutionary modeling that were not addressed within this work, and also points to possible directions for future research with the SEA:

- Investigations might be made regarding new evolutionary operators that would allow a variable number of spatial objects from generation to generation to exist. Those operators, tentatively named *death* and *birth* processes, would allow us to model more phenomena.
- Experiments into more complex planar location problems with continuous demand, variable organism size and shape could be designed.
- There is value in researching the modeling of fuzzy information using the SEA; fuzziness is an intrinsic—though frequently forgotten—property of spatial models.
- New applications could be developed using the SEA framework that are different from presented model. Some applications that potentially could be targeted for the SEA are:
  —maximum precipitation modeling;
  —redistricting and aggregation of spatial units;
  —suitability modeling;
  —gravitational models;
  —spatial interpolation.
- As evolutionary operators would be an attractive alternative to current modeling methods, there is merit in considering the integration of an evolutionary model into a commercial GIS package.
- Modeling with changing objective functions—as well as modeling for alternative solution (Pareto sets)—might be considered.
- The issue of two-level evolution, which was covered in this work at an operational level, might be investigated further, the key question being whether two-level evolution is merely a conceptual tool, or whether it really improves evolutionary algorithms.
- The issue of mating, using something other than distance functions—such as decay or extrapolative methods—should be addressed.
- The issue of spatial operators on indexed spatial data might be addressed—as this might improve the analysis of large data sets. Some of the questions that could be addressed here include: *How is the mating operator designed for indexed spatial data? What is a cross-over operator for index spatial data? How is the location index coded into genes? How are learning algorithms designed for indexed data?*
- The issue of the dynamically adjusted operators should be explored in an attempt to discover whether dynamic operators improve the performance of the SEA.

Table 2-8 Different EAs and data structures these algorithms use

Evolutionary Algorithm	Represented Problem or Class of Problems	Manipulated Data Structures	Operators
GP	Creation of computer programs	Trees, networks	Tree and network operators and algorithms
ES	Optimization of algebraic functions	Real numbers	Algebraic operators
EP	AI, automated rule generation	FSM, tables	Operators specific to FSM
CGA	Learning, optimization, classification	Binary strings	Binary string operators
SEA	Optimization of spatial systems and spatial representations, spatial reasoning	Spatial objects	Map algebra, spatial algorithms

SEA—a Parting View

Before we proceed to the last part of the book we would like to offer you a view of spatial evolutionary algorithms from the perspective of other evolutionary models. This view should help you to make up your mind (if you have not yet made it up) where in the universe of evolutionary algorithms the SEA fits (and why). The differentiating factor, we would like to argue, between SEA and diverse evolutionary algorithms relates to the problems these algorithms model and the data structures that these algorithms use. In table 2-8 we bring together in this context different classes of evolutionary algorithms. In this classification, spatial evolutionary algorithms stand out clearly as a separate class of evolutionary models dedicated to models of spatial phenomena operating on complex spatial objects.

After we read and reread the manuscript of this book we thought that few important points have not been clearly elaborated. As we did not want to disrupt the structure of the book already paid for dearly we decided to add some notes that would address our concerns.[17–20]

Notes

1. It is worth pondering whether a specific data structure (and with it the class of problems) that EA is designed to manipulate is the principal factor in differentiation of various classes of EAs. A look at the different types of EAs discussed in this chapter (EP, ES, CGA) shows that each of these algorithms has been specifically designed to manipulate a separate class of data structures. Of course, the choice of the data structure necessitates the design of genome and operators.

2. How do we apply the term "language" here? Illustrations from mathematics and music offer a convenient analogy. We refer to mathematics as a language describing abstract mathematical objects and relations. Similarly, we refer to musical notation as a code for recording a specific physical phenomenon that we perceive as sound. By analogy, we refer to spatial evolutionary models as a set of concepts for describing evolutionary models of spatial systems.

3. We would draw an analogy to the way in which spatial databases are regarded as distinctly different in their properties in comparison to nonspatial databases.

4. The quoted interpretation of the meaning of a system is due to Alexander (1979).

5. Interested readers seeking to enhance their understanding of spatial systems and spatial objects are directed to an exceptional book written by Alexander (1979). Although his book predates the development of object-oriented (OO) languages such as Java or C++, as well as the emergence of OO computer models, the unique contribution of Alexander's book is in its presentation of the concepts of objects and systems of spatial objects, which is done with a clarity and insight unmatched in contemporary OO literature.

6. The reader is directed to Mitchell (1996) for a fuller discussion of weak vs. strong modeling methods.

7. Dibble and Densham (1993) offer a similar critique of the results acquired by Hosage and Goodchild.

8. Tietz and Bart's algorithm is a generally accepted heuristic for the p-median modeling.

9. There is an analogy with United States Postal Service zip code areas.

10. The CCS is equivalent to 100 call-seconds. The *call-second* is a unit used to measure communications traffic (see note 11), equivalent to one call 1 second long. One user making two 75-second calls is equivalent to two users each making one 75-second call. Each case produces 150 call-seconds of traffic. 3600 call-seconds = 36 CCS = 1 call-hour. 3600 call-seconds per hour = 36 CCS per hour = 1 call-hour per hour = 1 erlang = 1 traffic unit.

11. The *erlang* is a dimensionless unit (*traffic unit*, defined above) of the average traffic intensity (occupancy) of a facility during a period of time, usually a busy hour. It has a value between 0 and 1, inclusive, and is expressed as the ratio of the time during which a facility is continuously or cumulatively occupied to the time that the facility is available for occupancy. *Communications traffic* (measured in erlangs for a period of time, and offered to a group of shared facilities, such as a trunk group) is equal to the average of the traffic intensity (in erlangs for the same period of time) of all individual sources, such as telephones, that share and are served exclusively by this group of facilities. (Source: FED-STD-1037C. Glossary of telecommunication terms. http://www.its.bldrdoc.gov/fs-1037/fs-1037c.htm.

12. In our case example, this assumption has been made.

13. At this point the reader may ask: What is the difference between the learning and mutation operators? At a certain level of abstraction, both mutation and learning affect the fitness of the individuals between the evolutionary cycles. Mutation may be thought of as randomized learning. The difference between those two operators is that mutation is a problem-generic functionality that is blind. That is, it is not guided by fitness. Mutation always changes a gene, based on some rule that is not related to the problem structure; sometimes mutation may be related to the evolutionary cycle. The learning operator is a problem-specific method that is specifically designed to improve the fitness of individuals. The learning operator, in our example, is based on a hill-climbing search. The latter is a method designed for covering problems. In other models, this operator may have a completely different form (while retaining its conceptual similarity to our learning operator).

14. "Two-level selection" is not a new concept to the theory of evolution. Lloyd writes: "There are number of cases in which selection is best described as several different levels and in which changes in a gene pool over time are best described in terms of an interaction of selection at suborganismic levels and selection at the organismic level" (Lloyd, 1993).

15. See the discussion of learning in evolution in part I.

16. In this analysis, one finding of particular interest was an observation that there was a very low number of cycles required for the model to converge. Indeed, earlier experiments with evolutionary programs have characteristically shown that evolutionary models require hundreds—if not thousands—of cycles for convergence to occur (DeJong, 1975; Mitchell, 1996). Such lengthy evolutionary cycles are consistent with what might reasonably be expected from an algorithm mimicking natural evolution (in nature it may take millions of years for new species to evolve). Consequently, to superficial inspection, the spatial evolutionary model with its short cycles, might appear as unexpectedly irregular and inconsistent in relation to knowledge of the operations of both mainstream evolutionary

algorithms and from paradigm of natural evolution. However, recent marine biological studies of fish species in Uganda's Lake Victoria (Anon., 1996) have revealed the finding that evolution of new species, in some cases, may require only a fraction (as little as one one-hundredth) of the time of that might normally be expected. For example, it has been shown that a new species of a fish discovered in Like Victoria (known as *cichlid*) is believed to have evolved in merely 20 generations. Such findings, applied to the presented model with its low number of cycles, would cast the latter as a special evolutionary case meriting further empirical investigation, and not an aberration.

17. It should be clear that whatever the data structure we claim to manipulate in our model, be it a tree, a network, or an FSM table, and whatever coding we claim to use, natural, binary or other, at the bottom of all this is an 8-bit word of 0/1 bits. A series of 0/1 bits is the representation that the real-world phenomena is encoded into and an 8-bit word of 0/1s is the data structure that our model manipulates.

18. To reinforce what we have said about genes and evolution we decided to quote verbatim the mini-reviews of two of Dawkins's books provided by Steward & Cohen (1997). Here are their comments:

> Richard Dawkins, *The Blind Watchmaker*, Longman, London 1986. [A beautifully written account of modern evolutionary theory—marred only by its basic premise that genes map to characters].
>
> Richard Dawkins, *The Selfish Gene*, Oxford University Press, Oxford, 1989. [Elegant account of neo-Darwinism: argues the view that DNA rules. Superb and well worth reading, but we don't believe it].

19. The approach to the spatial evolutionary modeling discussed in this book has been based on the idea of evolving a spatial system of objects by manipulating populations of spatial objects. This approach seems to be consistent with the design methodologies followed in other evolutionary models: ES manipulates FSM, GS manipulate tree structures, etc. However, a different approach to the spatial modeling is possible. It may be feasible to evolve systems of spatial objects not from the objects themselves but from the "bits-and-pieces" of space that constitute spatial objects. We could think of a particular spatial system as composed of atomic elements—tiles, lines, points, etc. By combining these elements into larger objects we could evolve spatial constructs in the similar way as we evolve more complex objects (strings, numbers) from random "1"s and "0"s in CGA. We did not pursue this idea any further here.

20. Research on new modeling methods (such as evolutionary programming, neural networks, data mining and knowledge discovery) in spatial information sciences should emphasize the specificity of spatial phenomena and concentrate on extending nonspatial methods to handle spatial problems. Thus the question guiding this research should be not how to model spatial phenomena with a particular modeling method but how this method can be extended to model spatial phenomena.

References

Al-Attar, A. 1994. A hybrid GA-heuristic search strategy. AI Expert, 10, 34–37.

Alexander, C. 1979. The Timeless Way of Building. New York: Oxford University Press.

Anon. 1996. Fast origin of species took just 12,000 years. The New York Times, August 1996, C1.

Armstrong, M. & D. Bennett. 1990. A bit-mapped classifier for ground water quality assessment. Computer and Geosciences, 11, 811–832.

Bailey, M.P. 1992. Measuring performance of integrated air defense networks using stochastic networks. Operations Research, 40(4), 647–660.

Bennett, D.A., M.P. Armstrong, & G.A. Wade. 1996. Agent mediated consensus-building for environmental problems: A genetic algorithm approach. In: 3rd International

Conference/Workshop on Integrating GIS and Environmental Modeling, Santa Fe, New Mexico.

Bianchi, G. & R.L. Church. 1992. A non-binary encoded genetic algorithm for a facility location problem. Unpublished paper.

Boucher, N.J. 1992. The Cellular Radio Handbook, 2nd Ed. Denver: Quantum Publishing.

Brocken, F.W.A. & B.J.M. Strelder. 1990. Mobile networks: planning and traffic distribution generation. In: L. Lada (ed.), Network Planning in the 1990's, 69–76. Elsevier Science.

Brooks, C.J. 1996. A genetic algorithm for locating optimal sites on raster suitability maps. First International Conference on GeoComputation, Leeds, U.K.

Chan, G.K. 1991. Propagation and coverage prediction for cellular radio systems. IEEE Transactions on Vehicular Technology, 40(4), 665–670.

Church, R.L. 1984. The planar maximal covering location problem. Journal of Regional Science, 24(2), 185–201.

Cooley, R.E. & M.H.W. Hobbs. 1992. An application of AI to computing class partition values for thematic maps. Proceedings of 5th SDH, Charleston, South Carolina, 371–380.

DeJong, K.A. 1975. Analysis of the behavior of a class of genetic adaptive systems. Unpublished doctoral dissertation, University of Michigan.

Delahaye, D., J-M. Alliot, M. Schoenauer, & J-L. Farges. 1994. Genetic algorithms for partitioning air space. In: The 10th Conference on Artificial Intelligence for Applications, San Antonio, Texas, 291–297. Los Alamitos, Cal.: IEEE Computer Society Press.

Dibble, C. 1994. Beyond data: handling spatial and analytical contexts with genetic based machine learning. In: T.C. Waugh & R.G. Healey (eds.), Sixth International Symposium on Spatial Data Handling, Edinburgh, U.K., 1041–1060.

Dibble, C. & P.J. Densham. 1993. Generating interesting alternatives in GIS and SDSS using genetic algorithms. GIS/LIS Proceedings 1993, Minneapolis.

Douglas, D. 1990. It makes me so CROSS, In: D.J. Peuquet & D.F. Marble (eds.) Introductory Readings in GIS, London: Taylor & Francis, 303–307.

Frank, H. & J.A. White. 1974. Facilities Location. Englewood Cliffs, N.J.: John Wiley.

Gamst, A., R. Beck, R. Simon, & E-G. Zinn. 1985. An integrated approach to cellular radio network planning. Proceedings of the 35th IEEE Vehicular Technology Conference, Boulder, Co, 21–25.

Goodchild, M.F. 1984. ILACS: A location-allocation model for retail site selection. Journal of Retailing, 60(1), 84–100.

Herring, J. 1991. The mathematical modelling of spatial and non-spatial information in Geographic Information Systems. In: D.M. Mark & A.U. Frank (eds.), Cognitive and Linguistic Aspects of Geographic Space, 313–350, NATO ASI series. Dordrecht, The Netherlands: Kluwer Academic.

Hobbs, K. 1995. Spatial clustering with a genetic algorithm. In: GIS Research U.K., Third National Conference, Newcastle, U.K., 20–30.

Hobbs, M.H.W. 1993. A genetic algorithm for knowledge discovery in spatial data. In: GIS Research U.K., First National Conference, Keele, 18–20.

Hobbs, M.H.W. 1994. Analysis of a retail branch network: a problem of catchment areas. In: GIS Research U.K. 1994, Leicester.

Holland, J.H. 1993. Adaptation in Natural and Artificial Systems, 2nd Ed. Cambridge Mass.: MIT Press.

Hosage, C.M. & M.G. Goodchild. 1986. Discrete space location-allocation solutions from genetic algorithms. Annals of Operations Research, 35–46.

Juliff, K. 1994. A multi-chromosome genetic algorithm for palette loading. In: Proceedings of the Fifth International Conference on Genetic Algorithms, 467–473. San Mateo, Cal.: Morgan Kaufmann.

Keller, S.F. 1995a. Potentials and limitations of Artificial Intelligence techniques applied to generalization. In: J-C. Miller, R. Webel, J.P. Largange, and F. Salge (eds.), GIS and Generalization: Methodology and Practice. London: Taylor & Francis.

Keller, S.F. 1995b. Interactive parameter setting of line generalization operators using genetic algorithms. In: Proceedings of the 17th International Cartographic Conference (ICC'95), 3–6 September 1995, Barcelona, Spain.
Krzanowski, R.M. 1997. Evaluation of spatial evolutionary algorithm for spatial modeling. Unpublished doctoral dissertation. University of London.
Kurner, T., D.J. Cichon, & W. Weisbeck. 1993. Concepts and results for 3D digital terrain-based wave propagation model: an overview. IEEE Journal on Selected Areas in Communications, 11(7), 1002–1012.
Lee, W.C.Y. 1985. Mobile Communications Design Fundamentals. New York: Sams & Co.
Lin, J-L., B. Foote, S. Pulat, C-H. Chang, & J.Y. Cheung. 1993. Hybrid genetic algorithm for container packing in three dimensions. In: Proceedings of the Ninth Conference of AI for Applications, Los Alamitos: IEEE Computer Society Press.
Loomis, L.R. (ed.). 1943. Aristotle. On Man and Universe. New York: Gamercy Books.
Lopez, R. & K. Vlahodimitropulos. 1995. Adaptive prediction using Lee's Model in a non-homogenous environment. In: B.D. Woerner, T.S. Rappaport, & J.H. Reed (eds.), Wireless Personal Communications, 179–184. Boston, Mass.: Kluwer Academic Publishers.
Mitchell, M. 1996. An Introduction to Genetic Algorithms. Cambridge: MIT Press.
Molenaar, M. 1998. An Introduction to the Theory of Spatial Object Modelling for GIS. London: Taylor & Francis.
Ohlson, B. 1995. Geographic data from satellites for cellular network planning. Geo Info Systems, 6, 32–42.
Openshaw, S. 1988. Building an automated modeling system to explore the universe of spatial interaction models. Geographical Analysis, 20, 31–46.
Openshaw. S. 1992. Some suggestions concerning the development of artificial intelligence tools for spatial modeling and analysis in GIS. Annals of Regional Science, 26, 35–51.
Openshaw, S. 1995. Developing automated and smart spatial analysis exploration tools for GIS. The Statistician, 44, 3–16.
Openshaw, S. & T. Perree. 1996. User centered intelligent spatial analysis of point data. In: D. Parker (ed.), Innovations in GIS3, 119–139. London: Taylor & Francis.
Pargas, R.P. & R. Jain. 1993. A parallel stochastic optimization algorithm for solving 2D bin packing problem. In: Proceedings of 9th Conference on AI for Applications, Orlando, Florida, 18–25.
Pedrycz, W. (ed.) 1997. Fuzzy Evolutionary Computation. Boston, Mass.: Kluwer Academic Publishers.
Pereira, A.G., R.J. Peckam, & M.P. Antulus. 1996. Genet: A method to generate alternatives for facilities siting using genetic algorithms. In: EuroGIS96, Lisbon, Portugal.
Peucker, T.K. & N. Chrisman. 1975. Cartographic data structures, American Cartographer, 2, 55–69.
Raper, J.F. 2000. Multidimensional Geographic Information Science. London: Taylor & Francis.
Russell, B. 1948. Human Knowledge. London: Routledge.
Samet, H. 1990. The Design and Analysis of Spatial Data Structures. Reading, Mass.: Addison-Wesley.
Steward, I. & J. Cohen. 1997. Figments of Reality: The Evolution of the Curious Mind. Cambridge, U.K.: Cambridge University Press.
Tomlin, D.C. 1985. Map Algebra. Cambridge, Mass.: Harvard University Press.
Tomlin, D.C. 1990. Geographic Information Systems and Cartographic Modeling. Englewood Cliffs, N.J.: Prentice-Hall.
Tornqvist, G., S. Nordbeck, B. Rystedt, & P. Gould. 1971. Multiple Location Analysis. Lund Studies in Geography, No. 12.
Tuan, Y.-F. 1977. The Space and Place. Minneapolis: University of Minnesota Press.

Turton, T., S. Openshaw, & G. Diplock. 1997. A genetic programming approach to building new spatial models relevant to GIS. In: Z. Kemp (ed.), Innovations in GIS 4, 89–102. London: Taylor & Francis.
Whitehead, A.N. 1920. The Concept of Nature. Ann Arbor, Mich.: Ann Arbor Books.
Worboys, M.F. 1995. GIS: A Computer Perspective. London: Taylor & Francis.

PART III

SPATIAL EVOLUTIONARY ALGORITHMS

Applications

In parts I and II of this book we introduced the nonspecialist reader to evolutionary algorithms and then outlined our approach to the design of spatial evolutionary algorithms. We have argued that a specifically designed *spatial* evolutionary algorithm is required to overcome intrinsic limits to the generic evolutionary algorithm. However, there are a number of different potential approaches to the design of a spatial approach and this part aims to present the best of the current approaches to this research problem.

Each of the chapters offers its own design perspective for spatial evolutionary algorithms. One of the strengths of the geographic information science field, but also one of its challenges, is the extreme diversity of the application areas from which problems spring. A wireless telecommunications application was discussed in part II and applications of spatial evolutionary algorithms to road design, airspace management, and landscape ecology are included among the following applications.

The chapters also spring from the need that geographic information scientists have for the manipulation of large and complex datasets with a wide variety of data types. Hence, the need for large-scale graph partitioning, the needs of multicriteria decision-making and the problems of patch optimization described in this part all defeat the use of deterministic algorithms and demand new methodologies, such as spatial evolutionary algorithms.

The applications are arranged from the most generic to the most specific. The first is presented by Catherine Dibble who proposes the development of genetics-based machine learning systems to examine complex arrays of spatiotemporally referenced information. The second application is presented by Chris Brooks who lays out a solution to the problem of the optimization of two-dimensional criteria-matching patterns in large search spaces. The third application presented by Steven van Dijk, Dirk Thierens, and Mark de Berg focuses on the complex problem of name placement on maps, which involves optimization of symbol placement, a hard computational task that cartographers, paradoxically, can do easily. The fourth application presented by Ângela Guimarães Pereira concerns the optimization of a multicriteria decision-making process for road path selection over complex terrains. The fifth and final application presented by Daniel Delahaye concerns the optimal partitioning of a three-dimensional airspace in order to minimize the crossing of aircraft passing through the space. The common factor uniting these applications is that they are all reaching towards ambitious new horizons marking computational solutions to problems embedded in space and time. The potential efficiencies in time and resource usage that solutions to these applications promise make this area of research both demanding and important.

3

Beyond Data

Handling Spatial and Analytical Contexts with Genetics-Based Machine Learning

CATHERINE DIBBLE

Geographic information systems (GISs) are fairly good at handling three types of *data*: locational, attribute, and topological. Recent work holds promise for adding temporal data to this list as well (e.g., see Langran, 1992). Yet the unprecedentedly vast resources of geographically referenced data continue to outstrip our ability to derive meaningful *information* from such databases, despite dramatic improvements in computer processing power, algorithm efficiency, and parallel processing. In part this is because such research has emphasized improvements in processing efficiency rather than effectiveness. We humans are slow-minded compared with our silicon inventions; yet our analytical capabilities remain far more powerful, primarily because we have evolved elaborate cognitive infrastructures devoted to ensuring that we leverage our limited processing power by focusing our attention on the events and information most likely to be *relevant*.

In GIS use, so far only human perception provides the requisite integration of spatial context, and human attention directs the determination of relevance and the selection of geographic features and related analyses. Understanding of spatial context and analytical purpose exists only in the minds of humans working with the GIS or viewing the displays and maps created by such operations. We still extract information from our geographic data systems primarily through long series of relatively tedious and complex spatial operations, performed—or at least explicitly preprogrammed—by a human, in order to derive each answer.

Human integration of analytical purpose and spatial and attribute contexts is perhaps the most essential and yet the most invisible component of any geographic analysis, yet it is also perhaps the most fundamental missing link in any GIS. Only humans can glance at a map of a toxic waste dumps next to school yards, or oil spills upstream from fisheries, and recognize the potential threat of such proximity; human cartographers understand the importance of emphasizing either road or stream networks depending on the purpose of a map; humans understand that "near" operates at different scales for corner stores versus cities, or tropical jungle habitat versus open savannah. Given a GIS with the capability to deluge any inquiry with myriad layers of extraneous data, this natural human ability to filter data and manipulate only the meaningful elements is essential.

Yet a GIS requires far richer information and knowledge structures in order to address such challenges directly. The importance of representing geographic meaning (Nyerges, 1991), context, and purpose (Couclelis, 1992) has been widely recognized, but practical application of these ideas has been hindered by the extreme difficulty of knowledge elicitation from experts (Lanter, 1992; Nyerges, 1991), by the difficulty of finding efficient representations for spatial relations within adaptive systems (Armstrong, 1991; Whigham et al., 1992), and less severely by the restrictive formal logic requirements of traditional forward or backward chaining inference systems (Smith et al., 1987; Lanter, 1992).

Genetics-based machine learning (GBML) systems encompass genetic algorithms, Holland classifiers, and genetic programming. Holland classifiers are adaptive rule-based expert systems that require neither prohibitively difficult knowledge elicitation and maintenance nor rigid formal logic structures. Classifiers have demonstrated their effectiveness in diverse practical aspatial problem domains (Goldberg, 1989; Holland, 1980, 1992a), but until now their potential for GIS has been limited by the lack of a general strategy for representing and evaluating spatial locations and geographic relationships within a GBML framework.

This chapter proposes such a strategy and discusses the use of genetic classifiers as frameworks for representing, storing, and (adaptively) accumulating information about meaningful subsets of space–time-attribute GIS data features given spatial context and analytical purpose. The next section, "Classifiers—Expert Systems that Evolve," introduces the traditional (aspatial) Holland classifier and its underlying genetic algorithm that serves as a research-and-development assistant in the development of new rules in the rule base. The section "Spatiotemporal Representations for Spatial Classifiers" proposes a spatial classifier that extends the traditional Holland classifier via an interface to GIS, tools for manipulating spatial data elements, and methods for evaluating spatial relationships within the classifier framework of rule and message strings. The last section, "Using Spatial Classifiers with GIS," introduces a simplified example of a spatial classifier for environmental monitoring and discusses design issues and potential applications.

Classifiers—Expert Systems that Evolve

The genetic classifier system is a cognitive architecture that allows the adaptive modification of a set of if–then rules. The architecture of the classifier system blends important features from the contemporary paradigms of artificial intelligence, connectionism, and machine learning, including

- the power, understandability, and convenience of if–then rules from expert systems,
- a connectionist-style allocation of credit that rewards specific rules when the system as a whole takes an external action that produces a reward, and

- the creative power and efficient search capability of the conventional genetic algorithm operating on fixed-length character strings. (Koza, 1992, 64)

Overall, induction acts continually and pervasively on a tremendous diversity of material. This gracefulness in accepting new information and goals, with little disruption of extant capabilities, comes close to being a hallmark of induction. Such gracefulness depends on the ready emergence of plausible, but tentative, knowledge structures integrating categories, relations, procedures, and expectations . . . (Holland et al., 1986, 346)

Holland classifiers (Holland, 1992b) are schema-driven (wildcard pattern matching), adaptive (via a genetic algorithm to breed the rules themselves) rule-based expert systems. What is "known" about the world is embedded in the rule base, and rules communicate with one another and with the outside world in the form of long character strings (messages). New rules are bred from the strongest (most useful) of the existing rules, and gradually prove their merit or are replaced by yet newer rules. Rules gather strength according to their share of eventual (possibly infrequent) payoffs to which they actively contributed. Such payoffs are passed back through the rule system by a bucket brigade algorithm, where each rule that was active at the time of the payoff passes back an appropriate share of the payoff to the rules(s) that sent the messages which activated it.

Genetic algorithms (Holland, 1992a; Goldberg, 1989) are computer search and optimization heuristics that use analogs to natural genetic evolution as heuristics for effective exploration of even extraordinarily complex problems (NP-complete). Potential solutions to a given problem are represented as character strings of fixed length. Genetic operators such as selection of the fittest, cross-over, and mutation are applied to successive generations of character strings to evolve desirable solutions.

Most of figure 3-1 illustrates the canonical structure of a Holland classifier (diagram adapted from Holland, 1992b). The "environment" section deviates from the traditional Holland classifier interface by including the specialized message types and the GIS data interface, which foreshadow the spatial classifier extensions that are presented in the section "Spatiotemporal Representations for Spatial Classifiers." The following annotated paragraphs briefly describe each of the classifier's components.

Message List

All communication to, from, and within a classifier is in the form of *message strings* stored in the *message list*. Each set of strings within a classifier is defined over a specific alphabet that is traditionally binary but may also contain integer characters. The representation of problem attributes and parameters for valid sets of message strings is central to the effectiveness of the classifier. The section "Spatiotemporal Representations for Spatial Classifiers" discusses the specialized roles of the *command* and *data* messages that constitute part of the spatial extensions.

Schema Patterns

Schemas have the same character positions as the message strings they are intended to match, but they add one more character to the alphabet, which serves

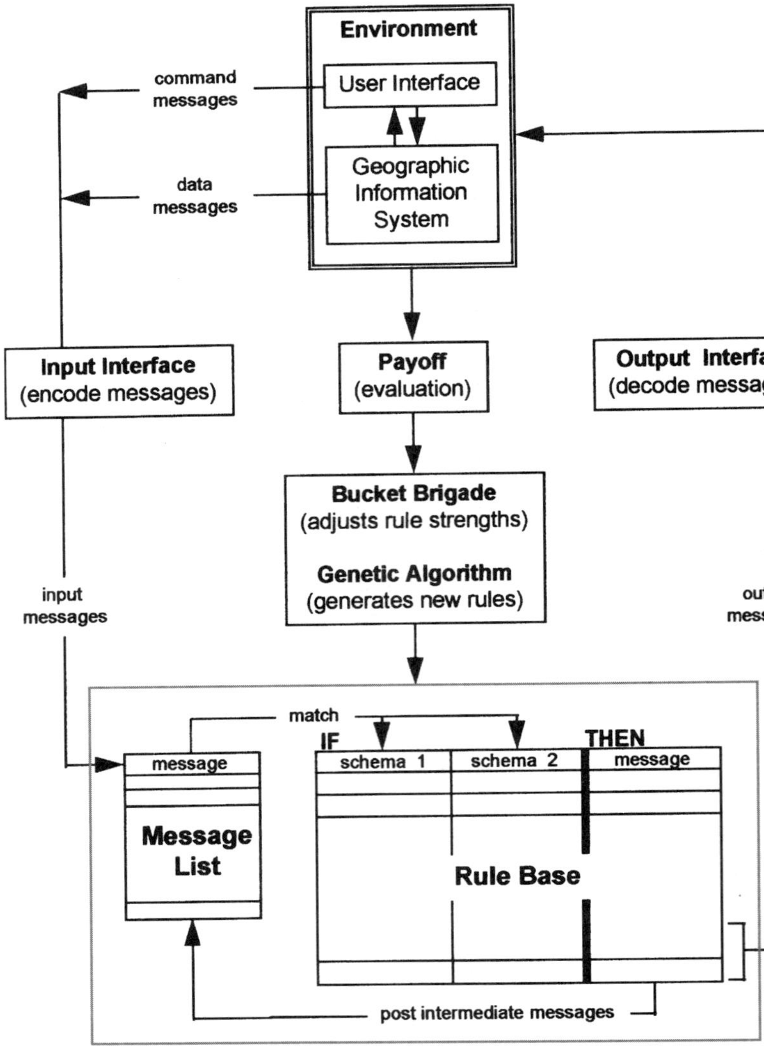

Figure 3-1 Holland classifier system. (Adapted from Holland, 1992b, p. 173.)

as a wildcard operator for matching patterns. For example, let * denote the wildcard operator, then the schema pattern 01*0 would match 0100 and 0110 in strings defined over a binary alphabet.

Input Interface

The *input interface* receives input from the outside world (via sensors or a human–computer interface) and encodes them into appropriate message strings for processing by the classifier.

Environment (Payoffs)

The classifier's *environment* is the real or artificial world against which the performance of the classifier is tested. *Payoffs* to the classifier are the cardinal representation of the evaluation of its success or failure in performing a particular task. For example, if the classifier is learning to produce a particular style of map, then the payoff could represent the cartographer's rating of the quality of that particular map. Note that payoffs can be relatively infrequent, occurring only after many cycles of the classifier and after many complex rule activations. Payoffs need not be completely objective as long as they are relatively consistent.

Genetic Algorithm to Breed New Rules

A genetic algorithm is applied to a particular decision problem by determining a representation of the parameters of the problem in the form of simple strings of parameters (the genes). These parameters traditionally take the form of bit strings of 1s and 0s, but can consist of other characters as long as the alphabet for each parameter remains quite small. The search for solutions begins by encoding a sufficiently large population of such parameter strings. In a traditional genetic algorithm, fitness for each string is measured by evaluating the parameters contained in that string. In a Holland classifier, strings represent rules and fitness is determined by the accumulated strength of each rule. New rules are derived from existing rules by applying three simple genetic operators:

1. *selection* of rules for reproduction, with probability proportional to the relative strength of each rule;
2. *cross-over* between parent rules to recombine promising portions; and
3. *mutation* of the occasional parameter to encourage continued variety.

Mutation rates are set to be very low, and play only a secondary role by ensuring some degree of novelty. Cross-over is instead the driving force in effective exploration, as above-average substrings of parameters are recombined during breeding of selected rules. Intuitively, we can think of the substrings as various building-block *ideas* about the best solutions to the problem under consideration. Thus, selection becomes a process of identifying rules that contain useful ideas, and cross-over then mixes and matches different combinations of good ideas as better and better combinations are discovered (Dibble & Densham, 1993).

The genetic algorithm within a classifier treats the rule base just as it would a normal population of strings, except that it breeds only a few new rules at a time, and is invoked only every *n* iterations of the classifier. The genetic algorithm uses each rule's accumulated strength as the measure of its fitness for selection.

Rule Base

The *rule* base of a classifier system differs from that of a traditional expert system in that the rules themselves have a more restricted form—expressed as character strings of fixed length—than the arbitrary rule structures that are usually explicitly programmed into expert systems. Each rule within a given classifier follows a particular form of a set number of concatenated condition schemas followed by an action message that is posted to the message list at time $t + 1$ if messages are found at time t that satisfy the rule's conditions.

For example:

$$\textit{if } \langle \text{schema 1} \rangle \textit{ and } \langle \text{schema 2} \rangle \textit{ then } \langle \text{message string} \rangle$$

where *if* and *then* are understood implicitly and *and* versus *or* for the conditions could be represented by a 1-bit flag at the beginning of the rule. Rules may take any such form, and in particular may have more complex conditions, but the rules must all follow the same pattern within any given classifier system.

Bucket Brigade

A rule at time t whose conditions have been satisfied competes, depending on its strength, for the right to post its message to the message list. This resolves ambiguity in cases where the messages from several activated rules would contradict one another. Rules that win this competition, and actually post their messages, in turn pay an appropriate internal strength fee to the rules that supplied the messages at time $t - 1$ that satisfied their schemas. Rules that happen to be active at the time of a payoff from the environment receive their payoffs from the environment and this outside "funding" provides the currency that is passed back through the bucket brigade credit assignment "information economy" such that rules which consistently contribute to high-payoff solutions gain proportionally in strength and thus are more likely to be "listened to" in the future. They are also more likely to be selected as the models from which the genetic algorithm breeds new rules.

Output Interface

Some of the messages posted by rules will contain special coding that indicates they are *action* or *output* messages. The *output interface* retrieves these messages from the message list and decodes them into commands that will be understood by the outside system.

Spatiotemporal Representations for Spatial Classifiers

Applications that rely purely on aspatial site-specific attribute information could be implemented within the traditional classifier framework that is described above and illustrated in the bottom section of figure 3-1. Yet it is inductive information structures for modeling *spatial* relations and context at geographic scales that would most enrich current GIS capabilities.

In particular, Holland classifiers require two new capabilities in order to handle spatial data gracefully. First, they need an efficient strategy for comparing and for learning relative distances and degrees of context-specific *nearness* between geographic features or grid cells. Yet explicit calculations of distances between all such items of interest would explode geometrically and prohibit attention to any but the most limited spatial relationships. Second, the messages within a given classifier need some way to *carry* specific pieces of geographic data for further processing such as display, aggregation, or more explicit analysis.[1]

I propose a general form of *spatial* classifiers, defined by three extensions to the traditional Holland classifier as it was described in the previous section:

1. a link to a repository of spatial data such as a GIS database (see figure 3-1);
2. extensions to allow efficient evaluation of spatiotemporal relationships; and
3. extensions to allow manipulation of specific spatial data elements.

In a spatial classifier, the usual encodings for attributes (theme) are supplemented by hierarchical encodings for space and time (Dutton, 1993; Langran, 1992; Nyerges, 1991; Sinton, 1978; Whigham, 1993). These are evaluated by two-stage prioritized comparisons of messages to rule condition schemas. The messages are divided into two separate classes that function as command messages and as data messages respectively.

A spatial classifier that learns about geographic structures and events in the full context of space, time, and attribute relationships must be able to make spatial and temporal comparisons very quickly *simply by comparing string representations of temporal and locational addresses*. Two key insights contribute to the design presented here. First, that hierarchical representations of space and time are powerful methods for representing a wide range of scales and precision (Dutton, 1993; Whigham, 1993). Second, that *nearness* (in either space or time) may be efficiently represented and tested by comparing hierarchical addresses for similarity at particular levels. Intuitively, if two locations are within the same cell of a hierarchical spatial tessellation, then they are considered to be "near" one another at that level. If they span a boundary at one level, they are likely to share the same cell address at the next higher level.

Command and Data Messages

Classifiers have traditionally had little to do with databases other than the knowledge structures that evolve within their own rule bases. Yet we need to provide our spatial classifiers with some means to know their world and to learn relevant attribute and spatial relations among the geographic features represented by a database. In addition, we need to provide the classifier with some way to refer to specific spatial items, whether a particular geographic location or feature.

For example, we might want the classifier to be able to learn about what happens to sensitive areas that are "near" marine oil spills. Then later we might want to ask it about places that were "near" the *Exxon Valdez* spill, which means that the input interface needs to have some way to encode both the nature and the purpose of the request ("find fish hatcheries near *this oil spill*") and then to provide the spatiotemporal data for the item ("*this oil spill* = location–time-attributes"). More generally, we need to provide the classifier with explicit means for encoding and communicating procedural knowledge (what to do[2]) and geographic data (what is where and what are its characteristics). The spatial classifier does this by providing two types of messages:

1. *command messages*: encode commands, actions, and internal communication among rules;
2. *data messages*: encode space–time-attribute data about database features or about a particular spatial feature in a query (what is near x at time y).

During classifier operation, command and data messages will generally circulate individually in the message list and be compared as usual with schemas for each rule condition. Under some special conditions the command messages need to be able to "pick up" a specific data message and carry it around for a while as part of performing some task, much the way that a worker ant might pick up a small piece of food to carry back to the nest. This can be accomplished with a simple toggle switch in the command message that effectively builds a tandem message linking the two for that time period.

Depending on the status of the "data-carrying" toggle of each command message, there can be three types of messages within the message list at any given time:

1. ⟨command string⟩
2. ⟨data string⟩
3. ⟨command string⟩⟨data string⟩

Extended Rule Structures

Rules in the spatial classifier can take many forms as long as they are valid character strings. The simplest rules would test for both a specific action or class of actions and one or more classes of data, which if activated would then post at least one command string, and possibly also an associated data string. As with the traditional genetic classifier, the *if* and *then* are understood implicitly, and the *and* or *or* are indicated by patterns within the command string schema. For example, these could take forms such as:

if ⟨command message 1 schema⟩ *and* ⟨data schema 1⟩

then

⟨command message 2⟩

or

if ⟨command message 1 schema⟩ *and* ⟨data schema1⟩ *and* ⟨data schema 2⟩

then

⟨command message 2⟩⟨data message that matches data schema 1⟩

Learning Spatiotemporal Relationships

Existing "spatial induction" systems have bypassed the problems inherent in generating spatial relations from raw data by pre-computing the relations that are believed to be relevant. This "de-spatializing," although useful with current induction systems, cannot uncover the spatial relations implicit in the data. This necessarily restricts the possible descriptions that may be generated, and is potentially a limitation when large amounts of spatial data, containing diverse spatial relationships, are involved. (Whigham et al., 1992, 402)

The central contribution of this chapter is a strategy for efficiently distinguishing general spatiotemporal nearness purely by string comparisons. This natural and efficient strategy complements the strengths of GBML systems and facilitates development of true *spatial* classifiers that inductively learn diverse spatial relationships in the full context of space, time, attributes, and analytical purpose. For simplicity, the discussion that follows will focus on representations and evaluations of space, and in particular of context-specific degrees of *nearness*, but the strategy is precisely the same for temporal evaluations except for the addition of a third-level comparison to indicate precedence in cases where that is relevant.

An essential insight here is the degree to which genetic classifiers depend for their learning on *general* patterns rather than absolute consistency. Thus, even if efficient representations capture important relationships such as nearness only *most* of the time, they can still provide powerful information for induction, *despite the cases that may be missed or happen to span boundaries.*

Space and time are represented as fixed-length character strings of hierarchical addresses, and may be either tessellations such as a quadtree or Dutton's quaternary triangular mesh[3] (Dutton, 1990, 1993; Goodchild and Yang, 1992), or other more subjective hierarchies that may have more natural interpretations for particular applications (Whigham, 1993). These addressing schemes and classifier rules are intended to apply to features in vector- or object-oriented databases or to individual or collected raster cells (depending on the level of the addressing scheme).[4]

There is one data message for each spatial object in an object-oriented database or one for each (set of) cell(s) of a raster system. Each three-part data message contains sections encoding the hierarchical temporal address of the feature,[5] the hierarchical spatial address, and the attributes related to the feature. This representation allows the spatial classifier to parse both absolute and relative space as follows (temporal neighborhoods are analogous).

Classifier alphabets are traditionally binary, but may more generally be represented by integer codings for specific attributes. The genetic algorithm's power to search the space of potential rules is greater if the alphabet remains relatively restricted, and thus it is wise to restrict encodings to limited integer sets for each parameter.

The data message alphabet may be defined for codes and addresses, where each attribute or address level is represented by small sets of characters chosen from the alphabets $\{0 \ldots 1\}$ or $\{0 \ldots 9\}$. The schema alphabet consists of all characters in the data message alphabet plus two wildcard operators that are parsed as follows:

* don't care at all—any data message character will match;

^ specified positions in two data messages must match one another exactly (but can have any value).

The ^ is the new wildcard operator that enables both relative and location-specific determination of degrees of nearness. In particular, it allows the spatial classifier to learn either global (applying anywhere in the database) or local (applying only in certain areas of the database) relevant degrees of nearness for particular patterns of attributes and purpose.

For example, using a modified QTM encoding (Goodchild & Yang, 1992), the spatial address schemas[6] 203023203^^^^******* for a particular rule to compare two data messages would detect matches among all data messages for features within Santa Barbara (the first nine exact digits, corresponding to a level ten resolution, or about 10 kilometers) that are also within a level 14 order of nearness (the next four ^ characters, to a resolution of approximately 500 meters) to one another. These spatial schemas indicate a database that can represent spatial resolution down to level 20, approximately 10 meters (Dutton, 1993).

Alternatively, the spatial address schema ^^^^^^^^^^^^^******* has thirteen "exact match" wildcards and detects matches for all data messages that indicate positions within approximately 500 meters of one another *anywhere* in the database, not just within Santa Barbara. Thus, the spatial classifier can develop rules that are sensitive not only to relative nearness but also to the influence of *location* within the database. *Near* for a grocery store may have very different connotations in Texas than in Boston. *Near* for a feeding range in a biogeography study may have very different meanings in the Kalahari desert than in a Congo jungle. This representation captures such distinctions in meaningful contexts that account for differences in attributes and analytical purpose (the command message) within a single rule.

For processing efficiency, message strings are compared to rule schemas in two stages:

Stage 1. Each position of each message in the message list is checked against each rule's schemas for everything except the spatial and temporal ^ (exact match) places.

Stage 2. After matches are found for all other conditions, then those (sets of) strings are compared with one another to find matches in the ^ positions.

A rule is considered to be activated once for each set of messages that satisfies both stage 1 and stage 2 matching for all of its condition schemas.

Using Spatial Classifiers with GIS

Spatial classifiers are most useful in complex situations where a wide range of attribute and spatial relationships affect the phenomena of interest. In particular, their comparative advantage lies in developing useful inductive data filters or models in situations that are too complex for representation and analysis by the more direct methods of logic or statistics. As such, no example simple enough to discuss briefly can capture the full range of an appropriate application for a spatial classifier. Nevertheless, this introduces one potential application and discusses classifier design and implementation issues.

Simple Example: Environmental Monitoring

> By improving the understanding of the relationships between stressors and landscape condition, these analyses should enable projections of relative risk to societal values given different landscape condition and stressor scenarios. (EPA, 1994, 19)

The United States Environmental Protection Agency (EPA) runs an Environmental Monitoring and Assessment Program that conducts research in landscape monitoring and assessment. Their three-step plan is to establish baseline landscape conditions from remote sensing imagery, to detect changes based on subsequent monitoring of the same areas, and to identify specific areas of significant change that may require more intensive monitoring and assessment (EPA, 1994, vii).

Six characteristics of this assessment program suggest consideration of a spatial classifier. First, the quantities of remotely sensed data to be analyzed are vast, while the salient data that reflect significant changes represent a relatively small subset that is difficult to identify a priori. Second, current monitoring plans are primarily concerned with detecting threshold changes in specific attribute values (EPA, 1994, 6). Third, any given site is expected to be evaluated only approximately every 10 years (EPA, 1994, 37). Fourth, environmental data are likely to be available from diverse sources and to be collected at disparate spatial and temporal resolutions (EPA, 1994, 24). Fifth, spatial patterns of environmental change may be evident only at specific scales (EPA, 1994, 17). Sixth, there exists a natural hierarchical framework of relationships both within and across landscapes and categories of risks (EPA, 1994, 2).

A spatial classifier could be designed to assist in the a priori identification of sites that are at high risk for environmental degradation and that would benefit from more intensive monitoring or intervention before threshold deterioration becomes evident. The classifier's data messages for each site would include:

1. the hierarchical QTM locational address;
2. hierarchical time stamp(s) associated with attribute values;
3. attributes describing physical site characteristic; and
4. attributes or indices describing monitored resources.

The classifier not only would evaluate data from sites monitored by remote sensing and pilot studies, but also could incorporate data messages from other sources that are related to geographic features of potential relevance to landscape conditions and categories of risks. The next subsection "Implementation Issues" discusses specific implementation issues in the context of this example.

Implementation Issues

Integration with GIS

Each spatial classifier would generally be implemented as a filter and coordination layer between a familiar GIS human computer interface (HCI) and the underlying GIS database and functions. Command messages are collected and encoded from the GIS HCI and appropriate data messages are encoded from the GIS database(s). It may be most efficient to store GIS data messages in a permanent file that serves as a form of quick reference index to geographic data. Classifier operation then filters the data messages in the index, and actual geographic data elements are retrieved only when mapping or other detailed analysis is identified as appropriate.

Evaluation Scores

Spatial classifiers learn and continue to refine their rules on the basis of repeated evaluations and associated payoffs from their environment. Payoffs need not be absolutely objective, but do need to be consistent and to be coded into cardinal values. For example, the spatial classifier for environmental risk assessment could be scored according to the success of its identification of high-risk sites. The classifier could receive a specific payoff for each site that is correctly identified as high risk. Each site used for training could be evaluated either objectively based on historical data and evidence of actual deterioration, or subjectively via assessments by human environmental experts.

Message/Rule Representation

As described in the section "Spatiotemporal Representations for Spatial Classifiers," hierarchical representations for locational, temporal, and even attribute data are ideal since they allow a spatial classifier to inductively discover the relevant patterns and relationships across various scales, rather than artificially restricting analyses to a uniform resolution. Classifiers learn most effectively when the alphabet for each character position remains very limited, either binary or with at most 10 valid characters. In addition, cross-over implies that specific string patterns consisting of positions that are close together are more likely to be preserved within the rule base. Hence, classifier effectiveness can be enhanced by coding locational and attribute information so that complementary combinations are relatively close together on each string. Effectiveness is further enhanced by ensuring that potential parameter patterns generated by cross-over are likely to generate valid rules rather than logical contradictions.

Priming the Rule Base

One of the greatest advantages of classifiers over other adaptive systems such as neural nets is the explicit accessibility and manipulability of their rule bases. A classifier can develop effectively from an initial rule base initialized purely from random generation of rules from the appropriate alphabets for each parameter. Yet a classifier may also be seeded with an initial configuration of potentially useful rules developed based on expert opinion or even merely human intuition about rules that seem likely to be helpful or relevant. Each rule then grows or fades in strength according to its usefulness as the classifier develops. Finally, a classifier's rule base is available for human inspection and modification at any time.

Modularity and Portability

At a system level, classifiers are naturally portable across specific GIS systems since all interaction with humans and with GIS data is already controlled by the input and output interfaces that encode and decode command and data strings as required. Such portability is less trivial at the data and application level, but can be facilitated by coding rules to be as general as possible with respect to data attributes (e.g., coding presence or absence of specific sets of characteristics rather than coding particular subjective attribute categories) and locational addresses (e.g. coding location using a hierarchical global addressing scheme such as QTM rather than a more localized quad tree or hierarchy of political units).

Related Applications

Geographic Modeling

Even sophisticated statistical models have limited ability to model the more intricate and convoluted relationships inherent in many areas of interest to geographers: a spatial classifier builds its models from "assemblages of synchronic and diachronic rules organized into default hierarchies and clustered into categories" (Holland et al., 1986, 342). A statistical model remains dependent on the insightful imagination of a statistician in order to specify and test the relationships among a particular set of attributes; it is essentially a process of testing potential *answers*, where each such hypothesis must be imagined and posed before it can be evaluated: a spatial classifier continually poses and evaluates a wide range of potential models. In addition, classifier systems are more forgiving with respect to their data quality requirements and can make graceful use of whatever data is relevant, despite differences in times, sources, resolutions, and attribute definitions.

Habitat Identification

A spatial classifier to identify elephant habitats could be fed data messages keyed by global QTM addresses and collected from several data sources. The key attribute would be the spatiotemporal specific observation of elephants in a particular area. The spatial classifier could be trained by asking it to learn to predict which areas would be likely to be favorable elephant habitats, and then scoring them according to comparisons between predicted and observed elephant ranges. In response, the classifier learns by building a model that pays attention to a possibly very wide range of human and natural geographic attributes that could be available in the database for the associated areas. This spatial classifier would be likely to notice *on its own* that elephant habitats in jungles need to be modeled differently from elephant habitats on the plains of the Kalahari, and would build in the appropriate locational sensitivities and intricate associations of other relevant attributes.

Acknowledgments Many thanks to Helen Couclelis and Mike Goodchild for their challenging questions and suggestions, and to David Mark and Geoff Dutton for advice on hierarchical structures. This research forms a portion of National Center for Geographic Information and Analysis (NCGIA) Research Initiative 8 "Formalizing Cartographic Knowledge"; funded by the National Science Foundation (SBR-88-10917).

Notes

The author began working with genetic algorithms in 1993 while at the Geography Department and National Center for Geographic Information and Analysis (NCGIA) at the State University of New York at Buffalo. See Dibble & Densham (1993).

1. This extension in particular may have fundamental GBML applications beyond GISs, since it essentially allows GBML classifier systems to act as intelligent interfaces for database processing.
2. *How* to carry out particular detailed procedural operations is most likely to be explicitly programmed within the GIS functions and simply controlled or called by the spatial classifier's output interface in response to specific command messages. Instead, the spatial classifier learns what to do and when and where to do it.
3. Quarternary triangular mesh (QTM) (Dutton, 1990, 1993) is a global hierarchical addressing scheme based on an imaginary octahedron embedded in the earth with points at the poles and at four equidistant points along the equator (at 0°, 90°, 180°, and 270°), where each face of the octahedron is then recursively subdivided into four "equilateral" triangles. Each address is identified by an initial 0–7 ID followed by hierarchical 0–3 IDs for each level of triangular subdivision.
4. Addresses could be defined only down to the appropriate level of resolution for their general scale. For example, the level 20 (10 meter resolution) QTM address for Waldo Tobler's house, per Goodchild & Yang (1992).
5. Of course, time can be omitted in applications where it is completely irrelevant. Also note that hierarchical addresses for time have the same advantages as hierarchical representations for space in terms of indicating the appropriate level of temporal precision or resolution.
6. The temporal and attribute sections are not shown here, for simplicity, but are an important part of the data message schemas for an actual spatial classifier.

References

Armstrong, M.P. 1991. Knowledge classification and organization. In: B.P. Buttenfield, & R.B. McMaster (eds.), Map Generalization: Making Rules for Knowledge Representation. London: Longman.

Couclelis, H. 1992. People manipulate objects (but cultivate fields): beyond the raster-vector debate in GIS. In: A.U. Frank, I. Campari, & U. Formentini, U. (eds.), Theories and Methods of Spatio-Temporal Reasoning in Geographic Space. Lecture Notes in Computer Science 639, Berlin: Springer-Verlag.

Dibble, C. & P.J. Densham. 1993. Generating interesting alternatives in GIS and SDSS using genetic algorithms. In: Proceedings GIS/LIS'93, 180–189.

Dutton, G. 1990. Locational properties of quaternary triangular meshes. In: Proceedings of the 4th International Symposium on Spatial Data Handling, 901–910.

Dutton, G. 1993. Toward more intelligent spatial data: reasons and rules for enriching locational notation. In: Proceedings NCGIA Initiative 8 Specialist Meeting "Formalizing Cartographic Knowledge," 24–27 October 1993, Buffalo, New York.

Environmental Protection Agency (EPA). 1994. Landscape Monitoring and Assessment Research Plan. U.S. EPA 620/R-94-009.

Goldberg, D.E. 1989. Genetic Algorithms in Search, Optimization, and Machine Learning. New York: Addison-Wesley.

Goodchild, M.F. & S. Yang. 1992. A hierarchical data structure for global geographic information systems. CVGIP—Graphical Models and Image Processing, January 1992, 54(1), 31–44.

Holland, J.H. 1980. Adaptive algorithms for discovering and using general patterns in growing knowledge bases. International Journal of Policy Analysis and Information Systems, 4(3); 245–268.

Holland, J.H. 1992a. Genetic algorithms. Scientific American, July, 66–72.

Holland, J.H. 1992b. Adaptation in Natural and Artificial Systems. Cambridge, Mass. MIT Press, 1975, revised and reprinted 1992.

Holland, J., K.J. Holyoak, R.E. Nisbett, & P.R. Thagard. 1986. Induction: Processes of Inference, Learning, and Discovery. Cambridge, Mass. MIT Press.

Koza, J.R. 1992. Genetic Programming: On the Programming of Computers by Means of Natural Selection. Cambridge, Mass.: MIT Press.

Langran, G. 1992. Time in Geographic Information Systems. New York: Taylor & Francis.

Lanter, D. 1992. Intelligent assistants for filling critical gaps in GIS: a research program. NCGIA Technical Report 92-4.

Nyerges, T.L. 1991. Representing geographical meaning. In: B.P. Buttenfield, & R.B. McMaster (eds.), Map Generalization: Making Rules for Knowledge Representation. London: Longman.

Sinton, D. 1978. The inherent structure of information as a constraint to analysis: mapped thematic data as a case study. In: G. Dutton (ed.), Harvard Papers on GIS, Vol. 7. Reading: Addison-Wesley.

Smith, T., D. Peuquet, S. Menon, & P. Agarwal. 1987. KBGIS II: a knowledge-based geographical information system. International Journal of Geographical Information Systems, 1(2), 149–172.

Whigham, P.A. 1993. Hierarchies of space and time. In: A.U. Frank & I. Campari (eds.), Spatial Information Theory: A Theoretical Basis for GIS. Lecture Notes in Computer Science 716, Berlin: Springer-Verlag.

Whigham, P.A., R.I. McKay, & J.R. Davis, 1992. Machine induction of geospatial knowledge. In: A.U. Frank, I. Campari, & U. Formentini (eds.), Theories and Methods of Spatio-Temporal Reasoning in Geographic Space. Lecture Notes in Computer Science 639, Berlin: Springer-Verlag.

4

A Genetic Algorithm to Design Optimal Patch Configurations using Raster Data Structures

CHRISTOPHER BROOKS

The design of optimal patch configurations is a generic problem relevant to many spatial planning exercises. Spatial pattern affects processes in the natural and manufacture of environment and should be incorporated as a criterion in planning. Currently, while geographic inormation systems (GISs) are adequate for data storage, analysis, and visualization they do not provide sophisticated spatial decision-making functions. With the help of GISs, pattern can be incorporated into spatial decision-making explicitly, using ad hoc procedures, or implicitly, through visualization of alternative plans. Other computer technologies like remote sensing and decision support systems facilitate decision-making by supplying timely data and techniques for solving multi-criteria evaluation problems. There are now a number of artificial intelligence techniques that can be coupled with GIS to address a variety of hard spatial problems. Genetic algorithms are particularly attractive for optimization problems because they are efficient and effective in complex search spaces.

Landscape ecologists use the twin concepts of patch and matrix to describe the spatial structure of the environment (McGarigal & Marks, 1994). The matrix is the dominant landscape element and patches are distributed within it. Patches can be crisp objects with well-defined boundaries, such as administrative areas, or inferred objects with fuzzy boundaries, such as vegetation or habitat patches in natural environments. In the former case, patches can be adequately represented by polygons in a vector GIS. In the second case patches are inferred from a continuous spatial distribution of attribute values. The raster data model is the most common representation of continuous fields within a GIS and is preferred to vector models in environmental applications because it is a better representation of the continuous variation characteristic of natural phenomena. There is a need for decision support tools that use raster GISs when spatial criteria relate to natural phenomena. Patch design involves many complexities: in a raster GIS it is also a complex problem in spatial geometry.

This chapter describes a genetic algorithm for designing patch configurations in raster GISs. The genetic algorithm is coupled with GIS and multicriteria evaluation functions to build an autonomous system that explicitly includes pattern as a criterion in the design of patch networks. Conceptually, the problem is to extract from an infinite set of possible spatial patterns a single pattern that is optimal by some criterion. The system must identify and evaluate alternative solutions, select the best and present a result that can be interpreted. Given a set of data inputs and a set of criteria for evaluating alternative patch configurations the algorithm is autonomous. The output from the genetic algorithm is a raster map in which each cell carries the identity of the patch to which it belongs. Matrix cells carry the value zero. The generated patches are not natural phenomena but are human constructs. The genetic algorithm output is a plan that must, depending on the application, be realized on the ground through some activity.

The next two sections describe the patch design problem. The first discusses how patch attributes affect utility and how patches can be measured and evaluated. This is followed by a section that describes the associated geometric problem in a raster GIS. Subsequent sections then describe the conceptual model for the genetic algorithm, a particular implementation, and a number of alternative possibilities. The key elements in the conceptual model are the representation of patches, the evaluation of solutions and the generation of better solutions using feedback information. The final section summarizes the chapter and gives some suggestions for further research.

The Patch Design Problem

Patches are characterized by their configuration and composition. Configuration is the spatial arrangement of patches: aspects of configuration include the shape of individual patches and the spatial relations among groups of patches. Composition describes the aspatial properties of a patch or landscape. Composition is a measure of the amount of different attribute values, while configuration is their spatial distribution. For a single patch, composition could be a single value denoting the dominant vegetation cover, land use or soil type, or it could be a vector of numbers denoting the amount of each type. Composition can also be measured using statistical summaries. It is important to note that, while patches are by definition considered homogeneous, by some criterion they will usually be inhomogeneous with respect to some attributes. A patch may be defined as suitable habitat for a species and is therefore homogeneous with respect to habitat suitability but will still include different vegetation types, soils, slope aspects, and so on. This is particularly true for designed patches, such as nature reserves. A nature reserve is homogeneous in the sense that all of it is reserve and it can be marked as such on a map, but on the ground it will incorporate variability.

The utility of a patch or landscape depends on both composition and configuration (Brooks, 1998). Both can be measured in a GIS and then evaluated using multicriteria evaluation. Composition is relatively easy to measure using a GIS by overlaying a patch map on to other attribute layers and extracting values. In a raster GIS each cell in the patch will return a value for each attribute. These values can be summarized in numerous ways such as minimum, maximum, or mean to give a single value for composition. Configuration can also be measured but it is a less precise concept. There are numerous indices for measuring the shape of patches and the pattern of patches including dominance, diversity and contagion (Forman & Godron, 1986); fractal dimension (With, 1994; Leduc et al., 1994); fragmentation index (Johnsson, 1995); proximity index (Gustafson & Parker, 1994); and the percolation index (Gustafson & Parker, 1992; O'Neill, 1992). Indices of shape and pattern may be difficult to evaluate with respect to specific objectives if there is no clear correlation with processes. However, patch configuration can be evaluated indirectly and usefully by measuring attributes that are themselves affected by configuration and that have a clear affect on processes. This is illustrated with the following examples from ecology.

In ecological systems, fragmentation and edge effects affect the effective area of patches, the interaction between patch and matrix, and the connectivity between patches. These all have significant influences on habitat suitability. The total area of habitat is not always of itself a significant measure. The same total area can be distributed in different ways that affect ecological functions. Small fragmented patches may be better than, or worse than, a single large patch, depending on the species concerned (Morrison et al., 1992). When small patches are of little value then the total area of patches greater than a threshold size is a more meaningful measure than the total size of all patches. Effective area may also be limited by edge effects. The edge effect can be defined as the distance d up to which influences form the matrix affect the patch (Laurance, 1991; Laurance & Yensen, 1991). Beyond a distance d from the boundary there is no outside influence. This zone is called the patch core. For core species, edge-effects limit the value of edge habitat. It may have no value whatever, in which case core area is the relevant measure. Other criteria relate to spatial pattern. There may be a requirement that patches are connected. Connectivity is a generic concept that can be evaluated in different ways; a simple way is to discount patches that are greater than a threshold distance from their nearest neighbor. As well as the spatial distribution of patches the shape of individual patches also matters. Shapes with long boundaries have greater interactions with the matrix and smaller core areas than more compact shapes. From these and other considerations it is apparent that different distributions of the same total area can all have different utility values depending on the relevant species, the ecological processes involved and the objective to be met. Counting effective area instead of total area is an example of an indirect way of measuring the effects of configuration. It is preferable to attempting to derive evaluation functions that relate indices of spatial pattern to utility because it is simpler and based on ecological principles.

The composition and configuration attribute values can be converted to a utility score for a patch design using multicriteria evaluation techniques. The use of multicriteria evaluation in GISs is well documented (Carver, 1991;

Eastman et al., 1993; Jankowski, 1995; Pereira & Duckstein, 1993). There are many techniques and they all involve procedures for comparing alternative solutions and selecting the best (Minch & Sanders, 1986). All techniques involve subjective judgments and no single method is ideal. The essential characteristic of multicriteria evaluation is that the criteria relate to disparate entities that cannot be measured on a single scale and therefore cannot be compared rigorously. There are two key difficulties. First, the criteria must be weighted to assign a relative importance for the objective. Second, measured attribute values have to be evaluated with respect to the criteria. This can be done using evaluation functions that convert the attribute values for each criterion to standard scores. Methods for defining evaluation functions are given by Pereira & Duckstein (1993). Standard scores are combined using a weighted summation, where the weight reflects the relative importance of the criterion.

The Geometric Problem in Rasters

Optimal patch design in a raster data structure is a large complex geometric problem. A patch is a cluster of contiguous cells and it is the contiguity constraint that makes patch design a difficult problem. Take the simple case of designing a single patch of N cells. This is orders of magnitude more complex than finding the best N cells without any contiguity constraints. Although the spatial constraint reduces the number of possible combinations of cells, the unconstrained, aspatial, problem can be solved using a simple efficient deterministic algorithm. The N-cell problem is a sorting problem. A simple algorithm to find the best N cells is to put N arbitrary cells in a group and compare every other cell against those in the group, rejecting the worst cell until all cells have been checked against the group. That would involve $N(S - N)$ comparisons, where S is the number of cells in the raster. This is actually much less than the number of possible combinations of N cells, which is $S!/[N!(S - N)!]$. The arbitrary patch problem is also orders of magnitude more complex than finding the best patch of a given shape, such as a circle. For a patch of a fixed shape, there are no more than S alternatives, one for each possible centroid of the shape. This is because there is only one way of arranging N cells into a circle at a given location. The following argument shows that for an arbitrary patch there are of the order $2^N S$ possibilities.

The position of a regular patch like a circle can be defined by a single cell location because the geometry is known. An arbitrary patch can also be defined by a single cell if the procedure for growing the patch is rigorously defined. Consider a simple process that starts with a seed and adds one cell at a time. The added cell must share an edge with the last cell added. Further, to ensure that the patches grown from any seed are unique to that seed, the new cell must have a higher row or column number than the last cell. This is illustrated in figure 4-1. Now, at each iteration there are two candidate cells so the number of alternative patches doubles. Thus the number of patches that can grow from a seed is 2^N. Each seed grows unique patches that cannot grow from another seed so the total number of patches is $2^N S$. Table 4-1 shows how the problem space becomes very large even for small values of N.

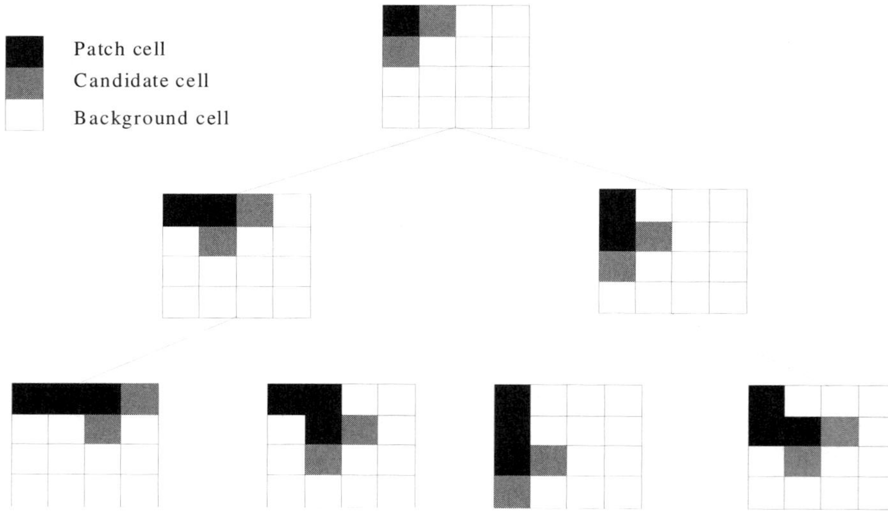

Figure 4-1 Alternative configurations of three-cell patches generated by the unique-patch generation algorithm: at each iteration there are two candidate cells, so in successive iterations the number of configurations doubles.

Complexity increases for multipatch problems and is even greater when the patches are of different types. The size of the problem dictates that it cannot be solved by complete enumeration even with massively powerful computers. Large problems, such as the N-cell problem, can often be solved algorithmically, so that size alone does not justify using a genetic algorithm. Here the genetic algorithm is used because there are no known deterministic algorithms. The problem is tackled using search and an efficient mechanism is needed to find the best solution in finite time.

The Conceptual Model

There are four logical components of the conceptual model: the search engine, the translation function, the spatial measurement functions, and the evaluation functions (figure 4-2). The genetic algorithm evolves a population of trial solutions

Table 4-1 Number of comparisons for different values of N: the patch problem rapidly becomes infeasible by complete enumeration

Number of Cells	N-cell Problem	Circle Problem	Patch Problem
5	4,975	1,000	32,000
10	9,900	1,000	1,000,000
15	14,775	1,000	32,000,000
20	19,600	1,000	1,000,000,000
25	24,375	1,000	32,000,000,000

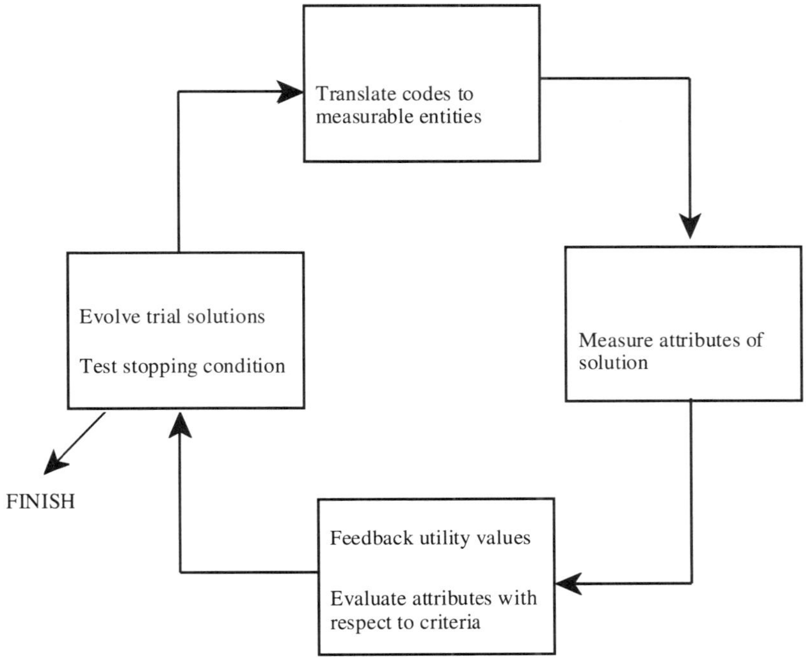

Figure 4-2 Conceptual model of the patch design algorithm.

until a stopping criterion is reached. Solutions are coded internally as genotypes that are translated into phenotypes, in this case raster maps. Spatial operators measure the composition and configuration of the patches and the measurements are converted to standard scores and then to a utility value by the multicriteria evaluation routines. The utility values for each alternative solution are returned to the genetic algorithm component as fitness values, which are used to bias the generation of a new population of solutions. When the stopping condition is reached the individual with the highest fitness is selected as the solution. It is presented as a map and therefore is meaningful to the end user.

The system takes two types of inputs, spatial attribute data and criteria information. Criteria are of two types, static and dynamic. Static criteria only relate to what is at a location and they do not change for different spatial patterns. Dynamic criteria relate to whole patches and patch networks and they can only be evaluated after a solution has been generated. Static criteria are be evaluated using the input data alone, whereas dynamic criteria can only be evaluated using the outputs. For example, the present vegetation cover attribute for a cell is an input and does not vary but the distance between two patches can only be computed when the patches are generated. Therefore, static criteria need only be evaluated once whereas dynamic criteria must be evaluated for every alternative solution. An example of the use of static criteria in GISs is suitability mapping. A habitat suitability index can be evaluated for each cell from attribute values to create a suitability map (Eastman et al., 1993). Such a map can then be input to the system as a single data layer. Dynamic criteria can relate to patch composition as well as interpatch relations. For example, one criterion may be patch diversity, which is a measure of the amount of different types of cell within the patch and can only be evaluated for a patch as a whole.

The Implementation: GAPD

Overview

The conceptual model has been implemented by loosely coupling a genetic algorithm search engine with a raster GIS and multicriteria evaluation routines. The system, called the genetic algorithm for optimal patch design (GAPD), uses the IDRISI GIS for the preprocessing and input of spatial data and for the storage and display of the output (Brookes, 1997b, 1998). The genetic algorithm routines were coded in C. For testing and evaluation the GIS measurement functions and multicriteria evaluation functions were also coded in C. The system runs on a PC or a UNIX workstation. Only basic measurement and evaluation functions were used and these were relatively easy to implement. Alternatively the system could have been built using library functions from systems such as GRASS (Baker & Cai, 1992) and FRAGSTATS (McGarigal & Marks, 1994).

A feature of the GAPD is that the translation function is performed by a parameterized region-growing program (PRG). The PRG is a heuristic for optimizing the configuration and composition of a single patch (Brookes, 1997a). Originally, the GAPD was conceived as a way of optimizing parameters to the PRG to overcome operational difficulties in using the PRG as a supervised tool. The PRG is more useful as a component of the GAPD where it fulfils the translation function. The PRG translates one-dimensional strings, which are a convenient structure for the genetic algorithm, into two-dimensional maps that can be evaluated to provide feedback to drive the search. The PRG grows patches under the control of a string of a parameters that define the location, size and ideal shape. The actual shape depends on the attribute values of cells in a number of input rasters. The relative importance of the ideal shape and the underlying data is controlled by a weighting factor where a weight of 1 means there is no deviation and 0 means that the ideal shape has no influence at all. The input rasters can be a single suitability map or multiple data maps.

The system operates in different modes depending on how much preprocessing is done on the spatial data. Figure 4-3 shows the case when the static criteria are preprocessed and a single suitability layer is input. Figure 4-4 shows the case when there are no static criteria and all criteria are handled dynamically within the evaluation routines. Inputs can be suitability maps, cost surfaces and raw data maps. The GAPD can be applied to single or multiobjective problems by designing appropriate objective functions (Brookes, 1998).

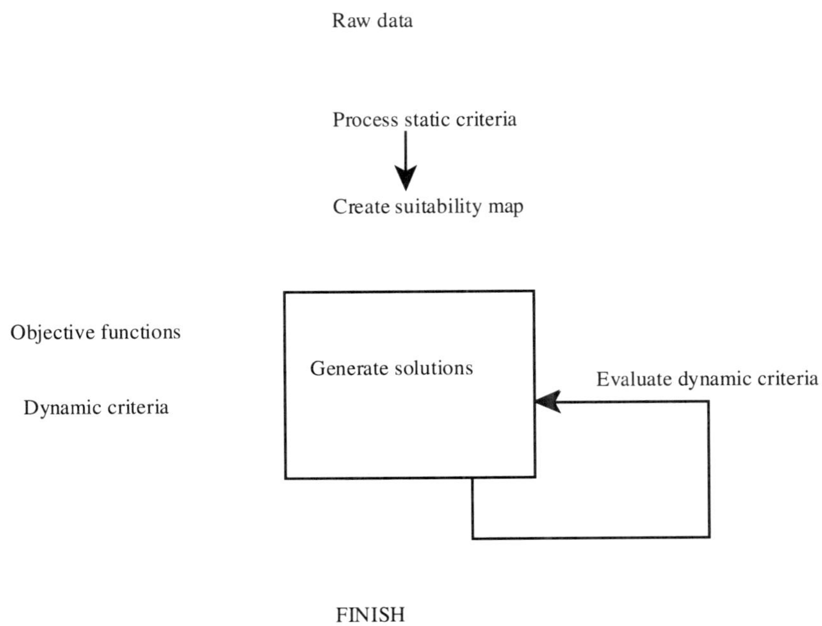

Figure 4-3 Static criteria can be preprocessed and input as a suitability map.

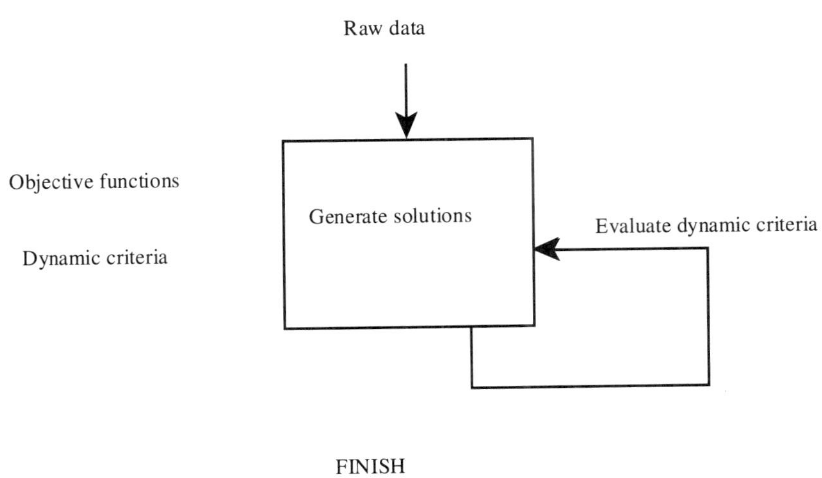

Figure 4-4 Operation when all the criteria are dynamic—raw data are input directly without preprocessing.

Genetic Data Structures

The genetic data structure is dictated by the PRG parameter set called a patch definition code (PDC). To control a single patch requires a location, a size, a shape, and a bias. The location is an *XY* coordinate pair giving the cell address, size is an integer number of cells, and the bias is a ratio. The basic model for shape control is anisotropic diffusion: the patch grows to a different extent in each direction. Three variables, constituting a shape definition code (SDC), control this process. The variables relate to the major and minor axes of the patch. An axis is a straight line from the seed to the perimeter. The longest axis is the major axis and the smallest is the minor axis. There may be more than one major or minor axis. The three parameters are the ratio, *R*, which sets the ratio between the minor and major axis; the orientation, *O*, which is the smallest angle between the *Y* axis of the raster and a major axis; and *N*, the number of major axes. A circle has an *R* value of 1. An ellipse has an *R* value less than 1 and an *N* value of 1. A large number of symmetric shapes can be generated (Brookes, 1997a). Asymmetric shapes can be generated by using multiple shape definition codes and combining the effects (Brookes, 1998). A simple PDC with one SDC is a list of seven numbers (six parameters—one is a coordinate pair) altogether and more complex versions have more.

More shape variety arises because of the influence of the underlying data. The PRG grows patches cell by cell starting at a seed and adding the highest scoring cell at each iteration. Each cell has two scores, a shape score, which depends on the SDC and the cell's position relative to the seed, and a cell score, which depends on the cell's attribute values and the cell scoring rules. The PRG can use different cell scoring rules depending on the application. In a simple case with one suitability map as the only data layer the cell score is read directly from the data. Other cases are possible; for example, if the objective is to optimize diversity in a patch, then the attribute value of a cell is compared to the current composition of the patch to assess how well the cell contributes to patch diversity. Thus, the PRG is not problem independent. This fact reflects the way the PRG was developed as a stand-alone heuristic, not as a component of the GAPD.

Multiple patch networks are specified by concatenating PDCs, one for each patch. The concatenated PDCs make a multipatch definition code or MPDC. In an MPDC each patch has an identifier and, optionally, a type to distinguish it from other patches. The identifier is implicit from the position of the PDC in the MPDC. If the patches can be of different types, then an explicit patch type parameter is added to the PDC. Within an MPDC the size parameter in each PDC is interpreted as relative size. The total size of all patches together is controlled by an additional parameter. The overall structure of an MPDC is a header, containing the combined size of all patches, and a tail, consisting of a string of PDCs. The relationship between PRG parameter sets, MPDCs, PDCs, and SDCs is shown in figure 4-5. Further details can be found in Brookes (1998).

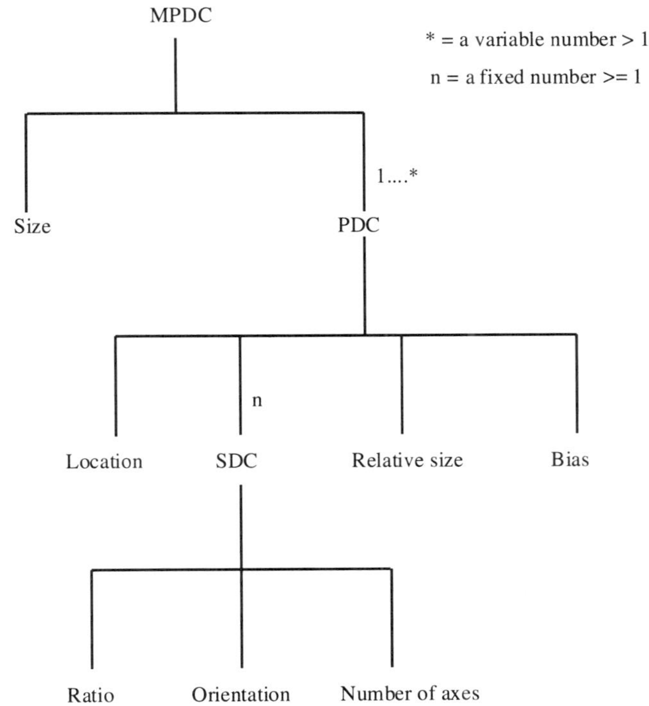

Figure 4-5 Hierarchical structure of the genetic code made up of MPDC, PDCs, and SDCs.

Genetic Operators

The genetic operators are of two types: the selection operators that create a breeding population from the current population and the reproduction operators that create new individuals from parents in the breeding population. Selection is a biased random process. The probability of an individual being selected is proportional to its fitness as computed by the evaluation process. Thus, fitter alternatives are better represented in the breeding population, their characteristics are passed on to the next generation, and the average fitness level tends to increase in succeeding generations. In each generation the best individual is copied unchanged into the next generation. This is simply to ensure that the best solution does not get lost, and doing so is justified because the objective is to solve a real optimization problem, not study genetics. The reproduction operators were designed specially for the numeric data structure and differ from the binary operators in the canonical genetic algorithm.

The GAPD uses cross-over and mutation operators similar to those in the canonical genetic algorithm but with one significant difference: the genes in GAPD are atomic, whereas in the canonical algorithm they are not. The numbers

11010011010000000

10011001000111111

Figure 4-6 Comparison of binary and GAPD cross-over. Binary cross-over can split genes and create new values but GAPD cross-over only redistributes genes without creating new values.

in the GAPD genetic code are MPDC parameters and each one is a gene that controls one aspect of the phenotype, for example, the number of axes. In the canonical genetic algorithm the genetic code is a binary string, genes are substrings, and the string elements are single binary digits. The binary cross-over operator swaps substrings between two individuals and the mutation operators modifies the value of a single binary digit. Thus, in the canonical genetic algorithm both cross-over and mutation can create new genes because the cross-over point can be anywhere in the genetic code, even within a gene. This is not the case with GAPD where the elements of the string and the genes are the same thing. The cross-over point cannot be within a gene, so cross-over swaps complete genes without modification. Cross-over generates new combinations of parameters but does not generate any new values. The difference between binary cross-over and GAPD cross-over is illustrated in figure 4-6. Mutation creates new genes by random changes to individual parameter values. Because the cross-over operator cannot generate new genes GAPD employs additional operators that each operate on a single gene. The three operators that are designed to increase the rate of genetic variation are *sum, average*, and *creep*. *Sum* and *average* operate on two parents and generate two offspring. With *sum* the two new genes are the sum and difference of the original values. With *average* the new values are different intermediate values between the two originals. *Creep* is like mutation but only changes the original value by a small value (e.g., 10 percent).

With mutation, *creep*, and *sum*, it is possible to generate invalid gene values. The mechanism to control this problem is the use of domains. Each gene has an associated domain, which is the range of valid values. Whenever a new value is generated it is checked against the domain and adjusted to a valid value. The domain is a torus and any overflow or underflow values simply loop around until they rest within the domain.

Operation

As well as the data and criteria there are a number of variables that control the operation of the GAPD. These include the probabilities for the different genetic operators, the population size, number of generations, and stopping criterion. Two stopping criteria are used: the GAPD stops either after a specified number of generations or when the average fitness has not increased for a specific number of generations. The GAPD handles constraints in different ways. Some hard constraints are handled using domains, for example, a constraint on total size is reflected in the total size domain. Other hard constraints, such as on maximum distance between patches, are handled by the evaluation functions. Soft constraints are handled using penalty functions. A penalty function reduces the utility of a solution by an amount relative to the extent to which it breaks a constraint (Goldberg, 1989). The penalty must be more than the benefit derived from breaking the constraint.

The implementation of the GAPD has demonstrated the feasibility of the concept. However, there are a number of issues concerning its operation. Criteria and objectives are coded as software routines and consequently the GAPD is recoded for each application. Ideally each logical component should be as problem independent as possible, so that only minimal changes are needed and thus making it easier to apply the GAPD to a wide range of situations. The genetic algorithm selection and reproduction procedures are independent of the problem as long as the PRG data structure does not change. However, because the PRG reads input data and grows different shapes according to different scoring systems it is not problem independent. The criteria and objective function are obviously determined by the problem. To some extent generic evaluation functions could be written and stored in a library but there will be no universal functions to solve all cases. The landscape ecology functions in the FRAGSTATS library (McGarigal & Marks 1994) are generic measurement functions and these could be extended or amended into a fairly comprehensive library. Standard methods for converting measurements to scores and scores to fitness values can also be implemented as library functions. The drawback resulting from PRG problem dependence can be addressed by using an alternative translation function.

Alternative Implementations

The same logical structure for the GAPD could be implemented in different ways. Currently the GAPD is a hybrid algorithm—the genetic search mechanism optimizes parameters to a heuristic region-growing program. In fact the GAPD was developed precisely to solve the problem of finding such parameters. Consequently, GAPD uses a particular mapping from strings to 2-D maps. One way to change the GAPD is to use an alternative mapping and an alternative genetic code. A problem independent genetic code and translation function would improve the usability of the GAPD by making it easier to implement for different situations. Another variation would be to keep the underlying PRG code but find alternative ways of coding multiple patches.

Currently the GAPD uses a concatenation of individual patch codes to define a network of patches. An alternative structure is a binary tree as used in genetic programming (Koza, 1992). In a binary tree each node is a patch and the whole structure represents a patch network. In figure 4-7 the left hand structure has five patches and the right-hand structure has three. The binary structure has a simple cross-over operator which exchanges nodes. In figure 4-7, cross-over has occurred at node 23 and node AB giving two new patch networks. The binary structure

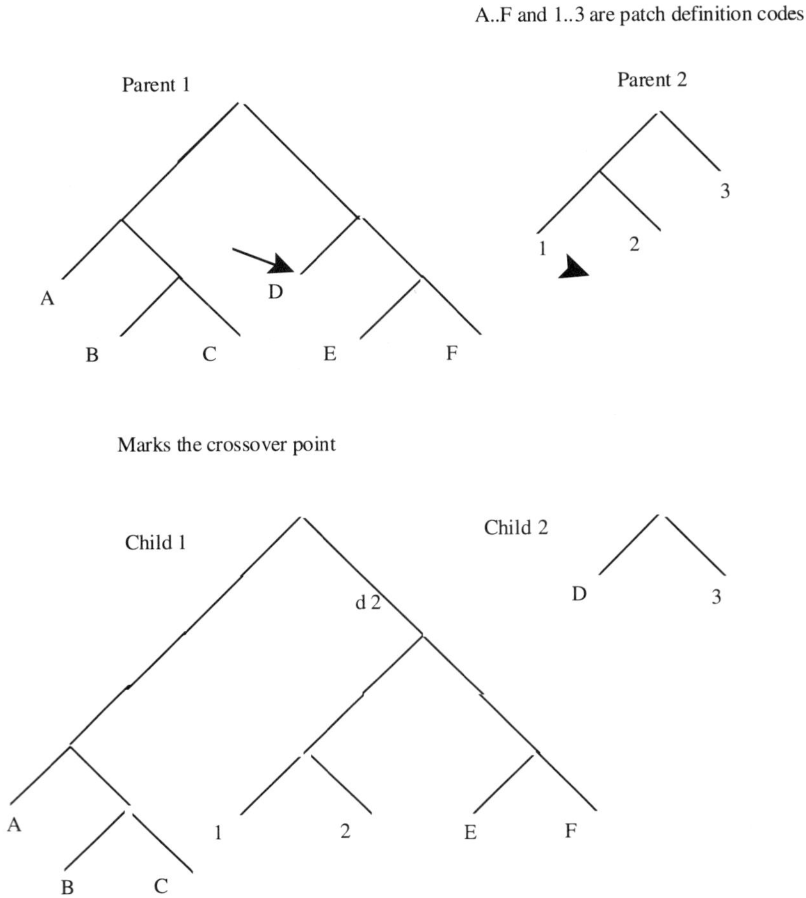

Figure 4-7 Binary tree genetic structure and cross-over operator.

gives a simple way of increasing the variety in patch networks but does not allow variation in individual patches. A second set of operators at patch level are needed to vary the characteristics of individual patches. These could be fundamentally the same as those used in the present linear structure. The potential benefit of the binary structure is an increased rate of variation in the patch network structure. Besides cross-over, possible tree operators include deleting nodes and adding new nodes.

Within the GAPD, the PRG is a means of coding 2-D rasters as 1-D strings. There are other ways of doing this that might be profitably exploited. Quadtrees are hierarchical spatial structures that can be used as spatial indexes or to compress data (Laurini & Thompson, 1992). A big advantage of quadtrees over the PRG is that they are domain independent. Another is that they can represent any spatial pattern, and can therefore generate any solution within the limits of the resolution of the raster, whereas it is not proven that the PRG can. As the spatial heterogeneity of a raster increases, the quadtree representation gets bigger so that there is a cutoff point where the quadtree gets bigger than the original raster. This is an important consideration for data compression. For designing patch networks it is a reasonable assumption that the patches are relatively large and relatively few in number and that the quadtree is an efficient data structure. Quadtrees can be coded as tree structures using nodes and pointers or as linear strings without pointers. In either case genetic operators can be adapted from other genetic algorithms or genetic programming methods. A quadtree-based GAPD would be a pure GA because it would not use another heuristic.

The advantages of quadtrees over the PRG are problem independence and the proven capacity to represent all possible solutions. One possible disadvantage is that potentially many solutions in the initial and early generations would exceed constraints, for example on the size of patches of a certain type, and thus be invalid. This may result in poor genetic algorithm performance because valid, but poor, solutions could swamp the population in early generations and lead to premature convergence on suboptimal solutions. There are mechanisms within genetic algorithms to address this kind of problem (Goldberg, 1989). One method is to use penalty functions that do not simply rule out invalid solutions that violate constraints but that give variable scores reflecting how far the constraints are broken. For example, a constraint on size that is only slightly exceeded has a smaller penalty than one that is greatly exceeded. This technique stops poor-quality solutions swamping the population because they do not have such a large fitness advantage as when invalid solutions have zero utility. The genetic algorithm should then avoid premature convergence and be able to evolve a population of high-quality valid solutions.

Applications

Patch design has applications in both natural and socio-economic contexts. Patch dynamics is very important in ecology and consequently in wildlife management and conservation. The design of protected areas and nature reserves must take into account factors such as disturbance regimes (O'Neill, 1992; Turner et al., 1993), minimum viable area (Gilpin & Soule, 1986), heterogeneity (Hanski & Thomas 1994), patch connectivity (Soule & Gilpin, 1991), patch-matrix interactions (Laurance, 1991) and climate change. These factors all contribute to the ways in which the size, shape and spatial distribution of patches affects ecological systems and their long term viability. Spatial factors also affect other natural phenomena, for example, the distribution of vegetation cover is important in fire spread and hydrological processes such as runoff. Managed forests are composed of stands of different age and type of trees and the pattern affects the economic return, wildlife value and recreation. In urban planning the size, shape and pattern of zones is important. For example, the position of residential areas relative to open spaces, amenities, commercial and industrial areas has a great bearing on their attractiveness as places to live.

Summary

Spatial pattern plays a crucial role in many natural processes and thus needs to be taken into account in many planning contexts. While computer systems exist for collecting, storing, and analyzing spatial data the functionality for design and decision making is inadequate. The GAPD is a successful implementation of a conceptual model for an autonomous system for designing patch networks. Configuration and composition are both optimized with respect to given objectives and subject to multiple criteria. The GAPD is loosely coupled with raster GISs and multicriteria evaluation tools and uses the PRG to translate 1-D structures to 2-D maps. Alternative implementations of the conceptual model are possible, including using a genetic code based on quadtrees. Further work is needed to develop the GAPD as a usable tool and to investigate issues in applying the GAPD to real problems.

References

Baker, W.L. & Y. Cai. 1992. The r.le programs for multiscale analysis of landscape structure using the GRASS geographical information system. Landscape Ecology, 7, 291–302.

Brookes, C.J. 1997a. A parameterized region-growing programme for site allocation on raster suitability maps. International Journal of Geographical Information Systems, 11, 375–396.

Brookes, C.J. 1997b. A genetic algorithm for locating optimal sites on raster suitability maps. Transactions in GIS, 2, 91–107.

Brookes, C.J. 1998. A genetic algorithm for designing optimal patch configurations in GIS. PhD thesis, University of London.

Carver, S.J. 1991. Integrating multi-criteria evaluation with geographical information systems. International Journal of Geographical Information Systems, 5, 321–339.

Eastman, J.R., P.A.K. Kyem, J. Toledano, & W. Jin. 1993. Explorations in Geographical Information Systems Technology, Vol. 4, GIS and Decision Making. Geneva: United Nations Institute for Training and Research.

Forman, R.T.T. & M. Godron. 1986. Landscape Ecology. London: John Wiley.

Gilpin, M.E. & M.E. Soule. 1986, Minimum viable populations: processes of species extinction. In: M.E. Soule, (ed.), Conservation Biology. The Science of Scarcity and Diversity, 19–34. Sunderland, Mass.: Sinauer.

Goldberg, D.E. 1989. Genetic Algorithms in Search, Optimization, and Machine Learning. Reading, Massachusetts: Addison-Wesley.

Gustafson, E.J. & G.R. Parker. 1992. Relationships between landcover proportion and indices of landscape spatial pattern. Landscape Ecology, 7, 101–110.

Gustafson, E.J. & G.R. Parker. 1994. Using an index of habitat patch proximity for landscape design. Landscape and Urban Planning, 29, 117–130.

Hanski, I. & C.D. Thomas. 1994. Metapopulation dynamics and conservation: a spatially explicit model applied to butterflies. Biological Conservation, 68, 167–180.

Jankowski, P. 1995. Integrating geographical information systems and multiple criteria decision-making methods. International Journal of Geographical Information Systems, 9, 251–273.

Johnsson, K. 1995. Fragmentation index as a region based GIS operator. International Journal of Geographical Information Systems, 9, 211–220.

Koza, J.R. 1992. Genetic Programming. Cambridge, Mass.: MIT.

Laurance, W.F. 1991. Edge effects in tropical forest fragments: Application of a model for the design of nature reserves. Biological Conservation, 57, 205–219.

Laurance, W.F. & E. Yensen. 1991. Predicting the impacts of edge effects in fragmented habitats. Biological Conservation, 55, 77–92.

Laurini, R. & D. Thompson. 1992. Fundamentals of Spatial Information Systems. London: Academic Press.

Leduc, A., Y.T. Prairie, & Y. Bergeron. 1994. Fractal dimension estimates of a fragmented landscape—sources of variability. Landscape Ecology, 9, 279–286.

McGarigal, K. & B.J. Marks. 1994. FRAGSTATS Spatial Pattern Analysis Program for Quantifying Landscape Structure. Forest Science Department, Oregon State University.

Minch, R.P. & G.L. Sanders. 1986. Computerised information systems supporting multicriteria decision making. Decision Sciences, 17, 395–413.

Morrison, M.L., B.G. Marcot, & R.W. Mannan. 1992. Wildlife habitat relationships. Wisconsin: University of Wisconsin.

O'Neill, R.V., R.H. Gardner, M.G. Turner, & W.H. Romme. 1992. Epidemiology theory and disturbance spread in landscapes. Landscape Ecology, 7, 19–26.

Pereira, J.M.C. & L. Duckstein. 1993. A multiple criteria decision-making approach to GIS-based land suitability evaluation. International Journal of Geographical Information Systems, 7, 407–424.

Soule, M.E. & M.E. Gilpin. 1991. The theory of wildlife corridor capability. In: D.A. Saunders & R.J. Hobbs (eds.), Nature Conservation 2: the Role of Corridors. Chipping Norton, NSW: Surrey Beatty & Sons.

Turner, M.G., W.H. Romme, R.H. Gardner, R.V. O'Neill, & T.K. Kratz. 1993. A revised concept of landscape equilibrium: Disturbance and stability on scaled landscapes. Landscape Ecology, 8, 213–227.

5

Designing Genetic Algorithms to Solve GIS Problems

STEVEN VAN DIJK
DIRK THIERENS
MARK DE BERG

What makes a problem hard for a genetic algorithm (GA)? How does one need to design a GA to solve a problem satisfactorily? How does the designer include domain knowledge in the GA? When is a GA suitable to use for solving a problem? These are all legitimate questions. This chapter will offer a view on genetic algorithms that stresses the role of the so-called *linkage*. Linkage relates to the fact that between the variables of the solution dependencies exist that cause a need to treat those variables as one "block," since the best setting of each individual variable can only be determined by looking at the other variables as well. The genes that represent these variables will then have to be transferred together. When these genes are set to their optimal values, they constitute a *building block*. Building blocks will be transferred as a whole during recombination and the building blocks of all the genes make up the optimal solution. As will become apparent, knowing the linkage of a building block is a big advantage and will allow one to design efficient GAs. Sadly, in the majority of problems, the linkage is unknown. This observation has given rise to a lot of development in linkage learning algorithms (for an example, see Kargupta 1996). However, there is a specific class of problems that allows for relatively easy determination of linkage: spatial problems. This is because in these problems, the linkage is geometrically defined. We will focus in this chapter on certain hard problems that arise in the context of geographical information systems and for which the linkage can be easily found. Specifically, we will fully detail the design of a GA for the problem of map labeling, which is an important problem in automated cartography. The map labeling problem for point features is to find a placement for the labels of a set of points such that the number of labels that do not intersect other labels is maximized.

The advantage of GIS problems is that the linkage of a building block is determined by the geometry of the problem. This allows us to make an efficient cross-over that does not disrupt good partial solutions. We also introduce a novel operator for explicitly creating building blocks (parts of the solution which need to be regarded as a whole): the geometrically local optimizer. This operator has several advantages, such as removing the need to tune a lot of parameters in the fitness function and making it relatively easy to incorporate new (e.g., aesthetic) constraints. We will give experimental results showing that the GA is indeed good at solving the combinatorial problem (maximizing the number of nonintersecting labels) but also that it is easy to extend the problem with additional demands.

This chapter is structured as follows. In the next section we will investigate the matter of linkage and building blocks further. Exactly what is meant by this, what is its importance and how can we determine it? We continue by reviewing problems that come up when using geographical information systems and we will see that these problems seem to be suitable for solving with a GA. However, we have to acknowledge the fact that GIS problems often include lots of additional constraints, which are difficult to cope with in a traditional GA. A way to solve this elegantly is then presented. We will put all these ideas to good use by solving the map labeling problem for point features with a GA. The performance of this GA is then compared against other algorithms. We then show that it is indeed easy to extend the problem with more constraints, and end with a discussion of our approach and a conclusion.

Genetic Algorithm Search

This section will introduce the notion of linkage and will show its relation to the difficulty of a problem.

Linkage

The reason some problems are hard and others are not is mainly due to interactions between the problem variables (a variable is coded in a GA by a gene). In an easy problem the variables have no significant higher order dependencies, which means that the best setting of each variable can be found independent of the others. Let's illustrate this by two problems, one easy (linear) and one hard. Both problems have three problem variables, which are defined on the binary domain. In other words, each variable can be either "1" or "0". The objective function for all combinations are shown in Table 5-1. A quick inspection of the *easy* function shows us that each time you flip a zero to one, you gain 10 points according to the objective function. This means that each x_i can be optimized in isolation and the concatenation of these optimal subsolutions gives the global solution (111). In this case it is possible to investigate the whole search space (all $2^3 = 8$ points in them). A GA is typically applied to very large search spaces and only has a few samples of it (the population). Based on those few samples it has to make educated guesses. Given the information $easy(000) = 0$ and $easy(001) = 10$ it has to deduce that it is worthwhile to search among solutions

Table 5-1

$x = x_1x_2x_3$	000	001	010	011	100	101	110	111
easy(x)	0	10	10	20	10	20	20	30
hard(x)	40	10	10	20	10	20	20	30

that have their rightmost bit set. This is done in an implicit manner, since on average, solutions with their rightmost bit set have a higher fitness and will therefore get selected more often.

Now consider the *hard* function, which is almost exactly the same as the *easy* one, with the exception that the global optimum now is located at 000. Considering the same samples and drawing the same conclusions, the search will progress towards the point 111, which is wrong. It is clear that the only way one can find the global optimum is by looking at triples of bits, instead of just one bit at a time. In this case it would mean that the whole search space would have to be searched. The three problem variables have a dependency between them that causes a need to regard them as a whole.

In general, when representing each variable as a gene, we want genes to be kept together when doing recombination if they represent variables with strong dependencies between them. This cohesion between the genes we call linkage. If there is linkage between three genes, the optimal setting for these genes is called a *building block*. The *building block hypothesis* says that the global optimum is found by combining (or *mixing*) building blocks. Mixing is the task of the cross-over operator.

Now that we now what linkage is, it is appropriate to ask how it relates to the hardness of a problem.

What Makes a Problem Hard?

The relation between the linkage between genes and problem hardness is provided by viewing genetic search as *statistical decision making* (see Goldberg, 1992). During the run of the algorithm we want the number of copies of a certain building block to increase until all the chromosomes have the settings of the building block for the genes that are involved. If this happens for the building blocks of all the other genes as well, we end up with optimal solutions. Let's take the example of the *hard* function again. It has three genes, which can be assigned values in 2^3 different ways. Of these 2^3 combinations, one is a building block (a combination which has the highest fitness contribution and which will fit in an optimal solution). The others are *competitors*. Suppose that instead of one building block, we have a problem consisting of many building blocks. If the recombination operator works correctly, the three genes will be kept together. How do we get the number of copies of the building block instead of some (suboptimal) competitor to be increased? Fortunately, the building block has an edge: by definition, if will have a slightly higher fitness contribution than a competitor. Therefore, on average, selection will favor chromosomes that contain the building block more than those that do not. To make this advantage significant, several factors are important. Especially interesting is the number of copies of the building block already present: if this number is large, the probability that selection (on average!) will favor the competitor will be small and the number of copies will increase.

If the initial population was randomly created by assigning random values to each gene of each chromosome, we can calculate how many copies of the building block are expected. Remember that for our three genes only one combination is a building block. Since all combinations are created with equal probability, the number of building blocks (for those genes) in the initial population is expected to be $n/2^3$ where n is the size of the population. If a building block would consist of four genes, the number of building blocks would be $n/2^4$. Since we need a certain number of copies to make the advantage of the building block significant, we need to enlarge the population. This causes longer running times. More linkage between genes can make the problem harder to solve. Suppose, however, that we know the linkage of the building blocks involved. We can then try to create building blocks ourselves and make the problem easier to solve. This is exactly what we will do when we design the GA for map labeling.

Implications

Problems are modeled as a number of problem variables that can be set to a finite number of values. Every combination of values is a point in the search space and the task of the GA is to find the point that is best according to some fitness function that takes the problem variables as input. Problem variables get coded as genes on a chromosome. A problem is hard when there are a lot of dependencies between variables: there exists linkage between the genes. The GA works by recombining groups of strong linkage that give a large fitness contribution: building blocks. These building blocks can be supplied in the initial population and they can also be created during the run of the algorithm. A GA works well if the building blocks get recombined (so they can form the optimal solution) and selection increases the proportion of building blocks in the population. To make an efficient GA for a problem of which we have a fairly good idea about the linkage of the building blocks, we can exploit that knowledge in two ways. Firstly, by making a cross-over operator that is linkage respecting in the sense that it transfers building blocks as a whole and does not disrupt them. Secondly, by making it easier for selection to increase the proportion of building blocks by explicitly creating them (using geometrically local optimizers, which will be described later). This leads to smaller population sizes and thus to shorter running times.

Let's try to answer the questions that started this chapter:

- What makes a problem hard for a genetic algorithm? A problem is hard when the building blocks consist of many genes, especially when we do not know the linkage of a building block. In order to find the best solution, the GA has to be supplied with an adequate supply of building blocks. If building blocks consist of many genes, the population size has to be very large.

- How does one need to design a GA to solve a problem satisfactorily? Besides ensuring a large enough population to allow the GA to find the building blocks, one has to design operators that preserve and repair them.
- How does the designer include domain knowledge in the GA? We will defer this question to the next section, where *geometrically local optimizers* are introduced as a way to repair and create building blocks and specifically guide the search into promising directions.
- When is a GA suitable to use for solving a problem? A GA is especially suitable for solving a problem when the linkage of the building blocks (which determines the structure of the problem) can be easily found or guessed. Since in spatial problems the dependencies of the variables are often related to the geometry of the problem (variables that depend on each other represent parts of the problem that are located close to each other), the linkage of the building blocks is easy to determine and GA's are a natural candidate to use for solving them.

In the next section we will look at a specific class of problems that have the advantage of allowing easy determination of linkage: problems that arise in GISs and that deal with spatial data.

Problems in Geographic Information Systems

In this section we will look at the kinds of problems that one can encounter in a geographical information system. A GIS deals with geographic data (points, polylines, areas, etc.) and their attributes (amount of rainfall, land use, etc.). Using a GIS one can process the data to show the information one is interested in. Showing the information can be done with graphs, but most often the map is the medium which is used to convey the information. Creating a map is a complicated process which was for centuries done by hand. In a GIS, this process has to be automated.

The kinds of problems that are interesting to solve with GAs are those dealing with large search spaces. This means that there have to be numerous variables which can be set in different states, and each combination of states can be more of less preferred.

For example, there is the problem of placing names on a map. Suppose the map is completely described by its geometric primitives (points, lines, areas). Each feature on the map (cities, rivers, etc.) has a name, which has to be placed on the map. A number of criteria apply to this name placement, the most important being that the name should not disturb other features or names much. This can become problematic when there is very little space. Also, the name should clearly associate with the feature it belongs to. The full fledged name-placement problem is quite complex, so let's begin with the point-feature name-placement problem (see figure 5-1).

A set of points is given and each point has a *label* (a rectangular shape on which the name will be printed). The label can be in one of four positions: on the

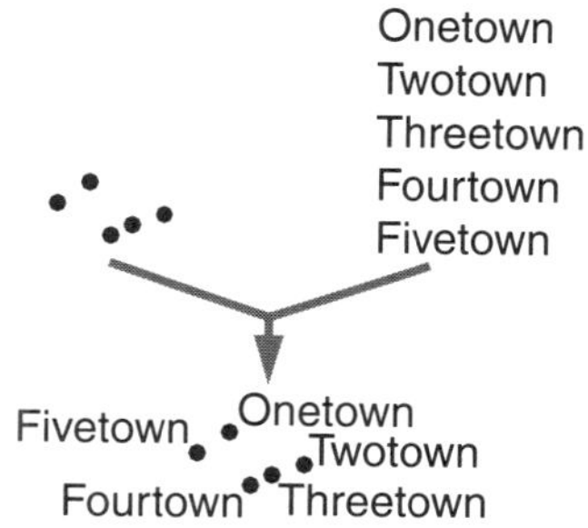

Figure 5-1 Point-feature label placement.

top right, top left, bottom right, or bottom left. An assignment of positions is required for all labels such that no two labels intersect. If this is impossible, one would like to maximize the number of free labels (i.e., labels that do not intersect any other labels). This problem is already NP-complete (for a proof, see Marks & Schieber, 1991), which basically means that trying to find a fast (deterministic, polynomial time) exact algorithm for finding the optimal solution is an infeasible approach. The only hope of finding a good solution in a reasonable amount of time is by searching the search space with a bias. That is what a GA does: it biases its search based on the information of the population, which gives it an idea where the good points in the search space are.

We solved the map-labeling problem using the GA, which will be described in the next section. It exploits the knowledge about dependencies between the variables, which in this case is caused by the neighborhood relation between cities. The placement of the label of a city is very much dependent on where its neighboring cities will place it. The farther a neighbor is from a certain point, the less it will influence the placement of its label. This determines the linkage of the building blocks and the whole GA is built around that.

Now consider the fact that someone who uses a GIS wants more than just to solve the rather abstract problem sketched above. First of all, he or she would like to consider more constraints than just the combinatorial aspects of the problem: perhaps consider the background of the map, delete certain labels but not the capitals, place labels in water for cities near the coast, or place the label in a preferred position whenever possible. How do we incorporate these constraints in the GA? A simple, but bad way to do it would be to somehow express these constraints in the fitness function. There are two major reasons why this is not a good idea. Firstly, most of the constraints are local and putting them in the fitness function gives a kind of global compromising effect that one does not want, like letting a label intersect because several others could then be placed in the sea. Of course one can try to balance these constraints by using weighting factors, to make sure that keeping a label free is more important than placing a label in a nice position. Ideally, the fitness function should only contain combinatorial constraints. Putting all constraints in the fitness function and using weighting factors to balance them is nevertheless often done. It brings us, however, to the second reason why putting additional constraints in the fitness function is a bad idea: all these weighting factors will have to be tuned. This will cost enormous amounts of time to run the GA on hopefully characteristic problems and it makes

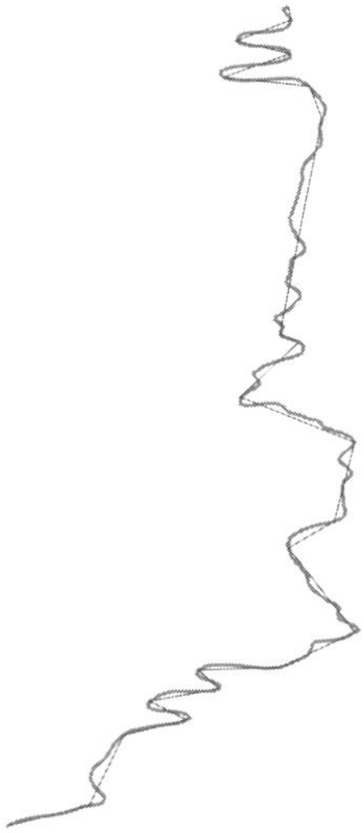

Figure 5-2 Line simplification.

the GA very inflexible, since adding another constraint will require a new tuning phase.

The alternative we propose is to use an operator called the *geometrically local optimizer*. This allows the new constraints to be handled outside the fitness function. It is applied to a local region of the map and locally optimizes the map. If it makes sure that enough alternatives remain, the GA will find the combinations of local regions that work well together (according to the combinatorial constraint of maximizing the amount of nonintersecting labels). This is also where the domain knowledge of the cartographer comes in, since on a local scale it is often relatively easy to say what is a good way to label a point.

Before we continue with the description of the GA for map labeling, it is instructive to show two other examples that were solved successfully with our approach. The examples illustrate the following points:

1. Solving the problem requires finding a good solution in a large search space.
2. The linkage of the problem is easy to determine.
3. Additional constraints need to be considered.

The first problem is to simplify a polyline (see figure 5-2). A sequence of points in the plane is given, which are connected by line segments. The task is to choose a subset of these points (including the first and the last one) such that the deviation of the resulting simplified line is within a certain error distance. Also, the new line is not allowed to intersect itself. Each point of the polyline can participate in the simplified line or not. If all the points are "on," the simplified line is the same as the original. If all points are "off" (except the first and the last point), the simplified line will be the straight line connecting the first and the last point. If the number of points in the input is n then the total number of search points in the search space is 2^{n-2}. A good simplification of the line consists of a number of long segments. Therefore building blocks probably consist of long segments, which we want to transfer as a whole during cross-over. Additional

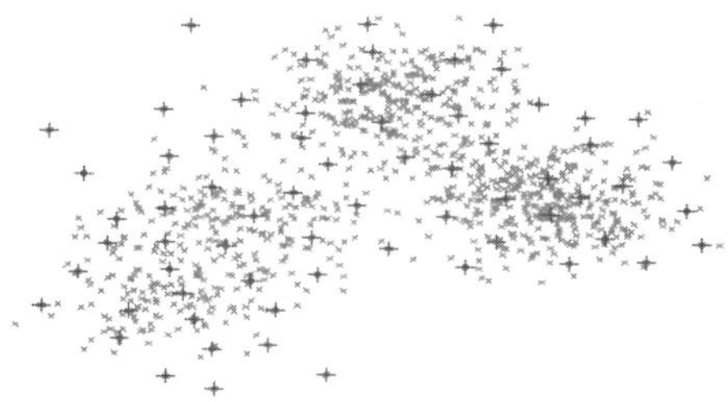

Figure 5-3 A generalization task.

constraints one may want to consider are ensuring that nearby points (e.g., border towns) stay on the same side of the line, demands on the shape of the line, or avoiding intersections when simplifying several lines simultaneously.

The second example (figure 5-3) is to select a minimal number of representatives among a set of points such that there is a minimal distance between two representatives and each point in the set is at a most some other distance away from the representative. Each point can be a representative or not, so the search space consists of 2^n possible solutions, of which some are invalid.

The linkage of the building blocks is again given by a neighborhood relation. An example of an additional constraint is ensuring that subsets of points contain at least one representative. Suppose for example that the points are shops of some food chain and the representatives give a picture of how shops are spread over a city. If the city is composed of certain districts, it is useful to make sure that each district that has a shop also has a representative.

A Genetic Algorithm for Map Labeling

This section will take the map labeling problem as a case study and develop a GA for solving it (also described in van Dijk et al., 1999). We will briefly discuss the various components to give a rough idea of the layout of the GA. After that, we will examine some of them in detail.

- *Fitness function*: the problem is to maximize the number of placed labels that do not intersect other labels (they are "free"), so we give each solution as the fitness the number of free labels. Obviously this function is to be maximized (ideally all labels will be free).
- *Encoding*: the encoding used will be an array of integers denoting position. If the map contains n cities, the array will hold n values between 1 and 4. We let 1, 2, 3, and 4 denote respectively the top-right, top-left, bottom-left, and bottom-right position (see figure 5-4).

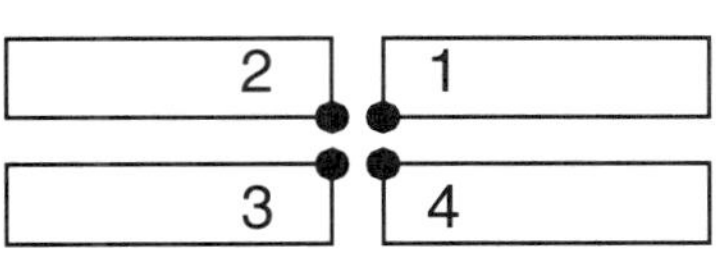

Figure 5-4 The possible positions of a label.

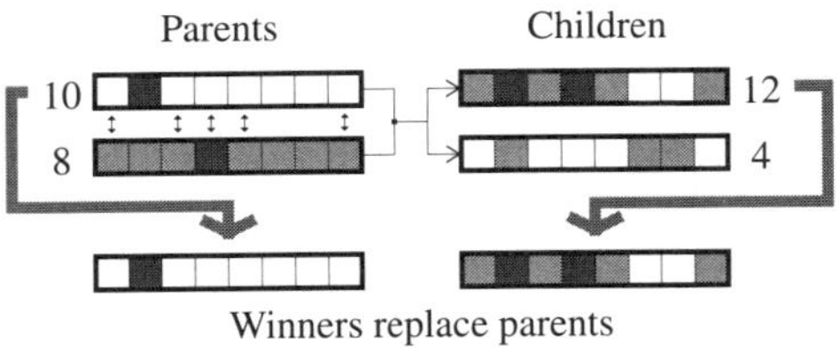

Figure 5-5 The elitist recombination scheme. Building blocks are dark gray.

- *Initialization*: for now, assume that we generate the first population of n individuals by randomly choosing a position for each label in each individual.
- *Update scheme*: with the update scheme we mean the way parents are selected and their offspring is placed back in the population. This GA will use an incremental scheme: only two parents get chosen at a time instead of a whole generation. The scheme used is the *elitist recombination scheme*.
- *Operators*: the cross-over operator recombines two parents but may introduce new conflicts on the resulting maps (new intersections between labels). To repair this (and perform other useful functions) the geometrically local optimizer is applied. No mutation operator is used.

First we will explain the update scheme in more detail, since this is the engine of the algorithm. It uses the operators cross-over and geometrically local optimizers to generate children with high fitness, so we will look at them in turn next.

Update Scheme

A genetic algorithm needs to evolve a population to find a good solution. Evolving a population requires an update step which is continuously iterated. For example, the population can be considered a generation which produces children that make up the new generation. This is a pure generation-based scheme. At the other extreme is the scheme which takes just two parents, instead of the whole population. This is called an incremental (or steady-state) scheme, which is what we will use.

The *elitist recombination scheme* (see Thierens et al., 1994) works by picking two parents at random from the population and evaluating their fitness. Then it applies the operators (cross-over and geometrically local optimizers) to generate two children and evaluates their fitness. From this family of four individuals the two individuals with the highest fitness replace the two parents. See figure 5-5 for an example. There one parent had such a high fitness that it did not get replaced, but the other parent was not so lucky. The selection pressure is due to this biased replacement. Note that in this example the number of building blocks in the population (the dark gray blocks) increases.

The advantages of this method are that good solutions will never be replaced in favor of worse solutions, it is conceptually simple and it will give good results if the operators are successful in making children with high fitness. Also, there is no need for a fitness scaling method, which is required with selection schemes such as proportionate selection. Instead the selection pressure is constant and similar to tournament selection (when the tournament size is 2).

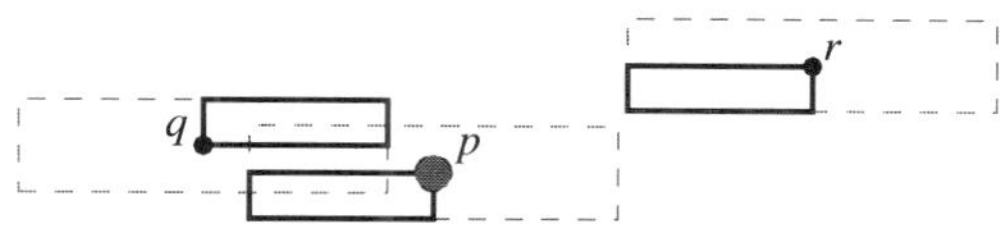

Figure 5-6 Points p and q are rivals, but p and r are not.

Cross-over

The cross-over operator needs to be carefully designed to keep variables (genes) that depend on each other together. In other words, the cross-over operator has to respect (potential) building blocks. So we can break down the design of the cross-over operator into two tasks: firstly, we need to determine the linkage of the building blocks; secondly, we have to transfer building blocks intact during cross-over.

Fortunately, in the map labeling problem it is relatively easy, at least on an intuitive level, to determine what the linkage is. It is clear that for a label it is very important how the labels of neighbors (points at a small distance away) are placed, and less (or not at all) important how labels of points far away are placed. To make this all less vague, we can define the concept of a *rival* (see figure 5-6), which is a special kind of neighbor. Not every neighbor is important, just the ones that can prevent the label from being placed somewhere. So we define two points to be rivals if they can place their labels in such a way that they intersect. If this is not the case, they are too far away from each other to be of direct influence (however, they can influence each other through rivals they both share, and so on).

Given a point p we can now define a *rival group* as p and all of its rivals. The goal of the cross-over operator should be to mix the genes from the parents, while keeping intact as many of the rival groups as possible. This done by accumulating rival groups until they make up about half of the map. More precise, we iterate the following procedure until the set S of points has size greater or equal than half the number of points on the map: pick a random point and place its rival group in the set S. We make children c_1 and c_2 from parents p_1 and p_2 as follows. The labelings of the points in set S are taken from parent p_1 and placed in child c_1. All the labelings for points not in S are taken from p_2. The other child is constructed likewise, but with the parents swapped (the labelings for points in S are taken from p_2 and so on).

This cross-over operator is successful in transferring building blocks (which we assume are rival groups) from parent to child, without disrupting them. However, since by the very definition of rival groups it follows that they overlap, there are also building blocks that can get disrupted during cross-over. See figure 5-7 for an example. There are two immediate ways of responding to this observation:

- Assume building blocks are larger (e.g., rival groups and *their* rivals). This is related to the question how much a labeling is influenced by the labeling of its rival's rival. This indeed eases disruption, but one needs to take care of *building block supply*. If the population is randomly generated one has to make sure that building blocks appear in it. For example, suppose we assume a rival group to be a building block. If a city has on average three rivals, a rival group of four cities will have $4^4 = 256$ different configurations it can be in, which can all appear in the population if it is generated randomly. A subset of those 256 configurations will be suitable for appearing in a optimal solution, and we

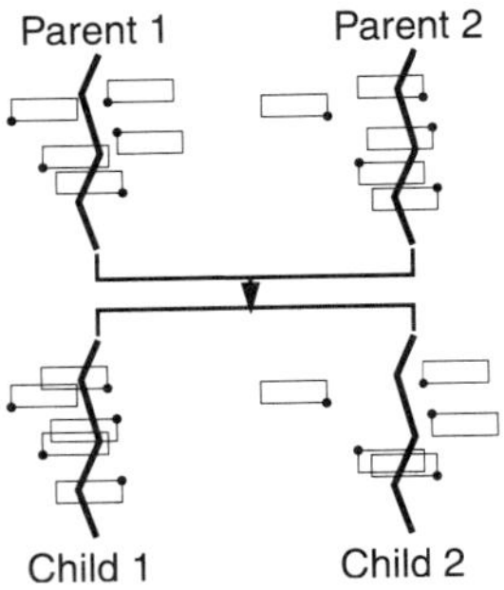

Figure 5-7 Cross-over can introduce new conflicts

have to make sure that enough copies are present in the initial population, so that the GA can filter them out. It follows that enlarging the size of your building block increases the demand on building block supply (since the number of configurations is larger, enough individuals have to be randomly generated to make sure the good ones appear in the initial population). This leads to larger population sizes.

- Let the stochastics of the GA handle it. Each building block has an above average fitness contribution (since it makes sure that some labels are placed freely). This contribution can be seen as a *signal* that has to be found among fitness contributions of other building blocks (called *collateral noise*). If the population size is larger, the probability of losing this signal becomes smaller.

Both solutions lead to larger population sizes, so there is clearly a tradeoff to be made. This is mainly dependent on the importance of the linkage between variables. There *is* linkage between a point and the rival of its rival, it is just weak (or more precisely: the dependencies between the variables the genes represent are small). One has to decide where to draw the line, and in this case taking a rival group as the building block is arguably best.

The above discussion is valid for all genetic algorithms that rely on statistical decision making for searching. However, in this particular case we have the benefit of knowing quite well how the linkage is defined among the variables. This gives us the opportunity to directly give the GA a helping hand. Let's start by battling disruption during cross-over, by visiting those points that may have a disrupted labeling and applying the operator from the next section: the geometrically local optimizer.

Geometrically Local Optimizers

As became apparent in the previous section, there is a need for an operator that can repair the conflicts which arising during cross-over. This operator is the geometrically local optimizer.[1] It is applied to a single point on the map and can only change the label of that point. We will first demonstrate how the geometrically local optimizer repairs the new conflicts. The use of the geometrically local optimizer extends beyond just repairing however. We will discuss the additional roles the geometrically local optimizer can play, such as the explicit generation of candidates for building blocks (which leads to smaller population sizes)

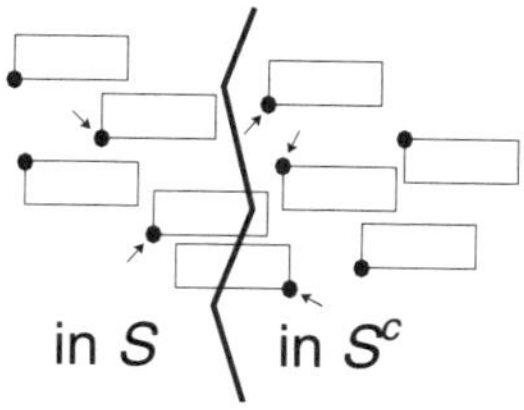

Figure 5-8 The points a geometrically local optimizer is applied to.

and the opportunity to extend the GA with new constraints and incorporate domain knowledge.

Consider the situation for one of the children after cross-over. Remember that S denotes the set of points that came from one parent. We will denote the set of points from the other parent with S^c. We can assume that a point from S that had its rivals all in S is without any worries. After all, the GA will make sure that it gets optimized through the normal mechanism of selection and mixing (which happens during cross-over). However, if a point from S has a rival that is in S^c (or a point from S^c has a rival that is in S), it is possible that it now has its label positioned such that it intersects with the label of the rival (see figure 5-8). We apply the geometrically local optimizer to all these points, as follows.

First we test the point for a conflict (an intersection with another label). If it has a conflict, the geometrically local optimizer tries to repair it: it determines for all positions (called *slots*) in which the label can be placed whether the label would intersect another label there. If it does not, the slot is called *free*. After determining the free slots, one is randomly picked and the label is moved to that position. This procedure is called *slot filling* (see figure 5-9).

A geometrically local optimizer should have the following properties:

- It should be able to improve, or at least not degrade, a geometrically local region on the map.
- It should not lower the fitness of the total solution, since that would cause selection to disfavor solutions that have geometrically local optimizers applied to them.
- When possible, it should choose randomly from multiple solutions that are equally good, to provide enough diversity for the global selection mechanism of the GA.
- It has to be fast, since it will be applied many times.

The role of the geometrically local optimizer is more profound than just repairing conflicts after cross-over. A better way to look at it is as an operator that explicitly tries to construct building blocks for the GA to work with. Whether the generated local solution is indeed a building block depends on the whether it will fit into an optimal solution. However, the normal mechanics of the GA will make sure that

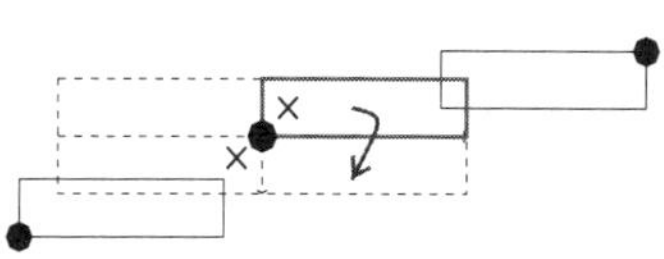

Figure 5-9 Slot filling.

these subsolutions get filtered out and the global solution is constructed. This is why a geometrically local optimizer should be careful to maintain diversity. However, just the fact that a geometrically local optimizer is applied in continually changing contexts will often ensure this without the need for any special alterations.

A geometrically local optimizer has many uses:

- It repairs conflicts after cross-over.
- It explicitly creates building blocks from which the GA can choose and construct a globally good solution.
- Since all constraints that are not combinatorial can be handled in the geometrically local optimizer, the fitness function becomes simpler. This means there is no need for a tuning phase, which would be time consuming, difficult to scale and make the GA inflexible.
- Extending the problem with new constraints: it now is possible to add constraints to the problem by handling them in the geometrically local optimizer. For a cartographer, it is often possible to say for a local situation whether something is good or bad. Therefore, it should be relatively easy to construct a geometrically local optimizer by allowing cartographers to express their expertise.
- Sometimes precedences in constraints have to be considered, which is possible using geometrically local optimizers but would be awkward to express in a fitness function. For example, for the map labeling problem there is a preferred order of label positions. Making sure a label does not intersect another label, however, takes precedence over considering a preferred label position.
- Pruning the search space: since the geometrically local optimizer only constructs local solutions that are good, a lot of bad search points of the search space become inaccessible. For example, if a point can place its label in four positions, but two will never be constructed by the geometrically local optimizer, the space that is searched is halved.

Note that geometrically local optimizers only get applied to points that can suffer conflicts after cross-over. If a lot of additional constraints are put in the geometrically local optimizer, the initializer that constructs the first population may need to be altered to make sure the geometrically local optimizer has been applied to each point on the map. This can be done simply by generating the population randomly as before, and then applying the geometrically local optimizer to each point of an individual in random order.

Robustness

Designing a genetic algorithm to solve a difficult problem can be challenging enough, but designing a GA for solving a GIS-problem poses some additional requirements. The intended users are likely to be completely untrained in the use of genetic algorithms, which means the GA should be constructed with as few algorithm parameters as possible, since every extra parameter is confusing. Secondly, for use in a GIS it should be possible to extend the algorithm without much problems. If the GA can live up to these requirements, we can say it is *robust*.

The GA that we sketched has the following points which account for its robustness:

- P_c, the cross-over probability, can safely be set to 1 because of the elitist recombination scheme. The cross-over probability has to balance between building block mixing (combining the good parts from the parents into a fitter child) and disruption (splitting something in the parent that was good). However, a child that is disrupted will become less fit and will not be able to replace one of the parents. Therefore there is no need to set P_c lower than 1.
- P_m, the mutation probability, can be set to 0 because geometrically local optimizers are used. Mutation is a mechanism that relies on chance to construct building blocks. Since geometrically local optimizers explicitly create building blocks, there is no need for mutation.
- No tuning of fitness weights is necessary because only the combinatorially hard aspect of the problem is measured in the fitness function.
- "Hitchhiking" is largely reduced because cross-over mixes on the level of building blocks. Hitchhiking occurs when a bad part of an individual gets transferred together with a good part to one of the children. It gets preserved only because the good part raises the fitness enough to let the child survive. Hitchhiking is a result of artificial linkage, caused by the cross-over operator. For example, with a one-point cross-over the assumption is made that building blocks consist of genes that are close together. If this is not the case and differences in fitness contributions of building blocks are large, hitchhiking will occur.
- "Genetic drift" is avoided because of the constant selection pressure that acts on all parts of the solution. Genetic drift is another peril a GA can become a victim of. If selection pressure on some of the variables is too low, the settings for those variables in the population will eventually converge to a single setting just because of stochastic effects caused by the fact that the population is finite. This happens, for example, when some building blocks have very large fitness contributions and others do not. Selection pressure on the latter will be low in comparison to the first.
- By applying the geometrically local optimizers, the search space gets pruned of search points that are clearly not optimal, which ensures the search space remains tractable.
- Extending the problem with new (noncombinatorial) constraints is easy, since the fitness function does not have to be altered.

Comparisons of the GA to Other Methods

Comparison for Map Labeling

We implemented the GA for map labeling and compared it against several other algorithms (see van Dijk et al., 1998) for more extensive results). In particular we considered the experimental study of Christensen et al. (1995) who compared several point-feature labeling algorithms and a paper of Verner et al. (1996) who described a GA for label placement. The study of Christensen et al. compared a 0/1 linear programming approach by Zoraster (1986), a greedy approach, a heuristic search approach by Hirsch (1982) and two algorithms of their own

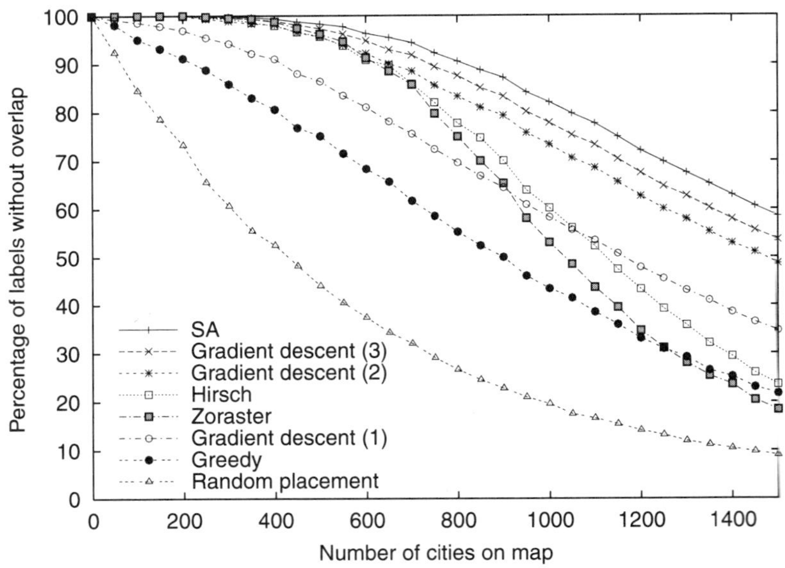

Figure 5-10 A comparison of label placement algorithms by Christensen et al.

devise: simulated annealing and gradient descent. Their conclusion was that simulated annealing achieved the best quality solutions. See figure 5-10 for their results.

Verner et al. designed a GA that used uniform cross-over and a masking technique to avoid disruption of good parts of the solution. They reported better performance than simulated annealing by using an eight-position model instead of a four-position model. We implemented simulated annealing, the algorithm by Verner et al., a hillclimber that randomly tried changes and kept only the improvements, and of course our own GA. We started by comparing our GA with the hillclimber. The hillclimber is a very simple algorithm and it is instructive to see how well it performs. Indeed, it performs rather well and is very fast. When problem instances get harder, it gets trapped in local optima. As expected, our GA outperformed the hillclimber (see figure 5-11).

More interesting was to see how the GA of Verner et al. compared against our GA. Unfortunately, we could not reproduce their results with our implementation. However, when comparing against the results from their paper, our GA turned out to perform better. Note that both GAs were run using the eight-position model. See figure 5-12 for the results.

The last algorithm, simulated annealing, turned out to have almost exactly the same performance as our GA. In figure 5-13 the results of this comparison are shown. Note that both algorithms used a four-position model, so the quality of the solutions is below the results of figure 5-12 where the algorithms used an eight-position model. Both algorithms (GA and SA) find solutions with almost exactly the same quality. A comparison between the speed of both algorithms showed that for relatively small maps (500 points and smaller) the GA was faster. For

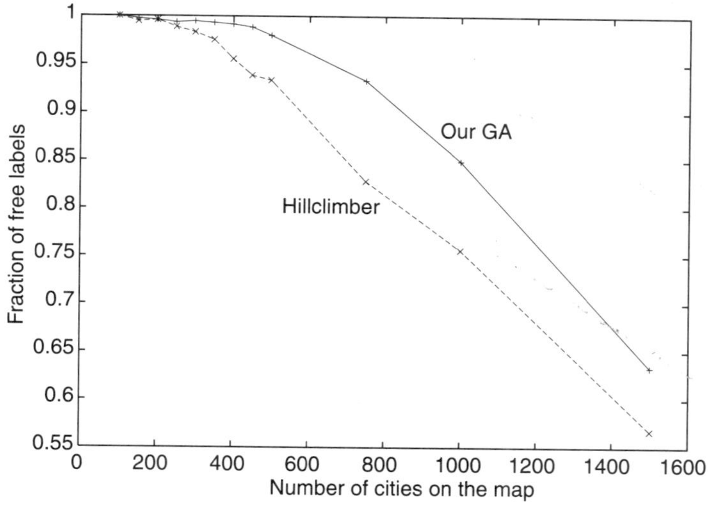

Figure 5-11 A comparison between our GA and a hillclimber.

bigger maps the SA turned out to be faster. However, we have not experimented with increasing the selection pressure yet, which would considerably speed up the convergence of the GA.

Summarizing, we can make the following conclusions:

- The hillclimber performed rather well, given its simplicity and the fact that it is very fast. However, for high-quality labelings more sophisticated algorithms are needed.
- The genetic algorithm of Verner et al. produces solutions with a lower quality than our GA.

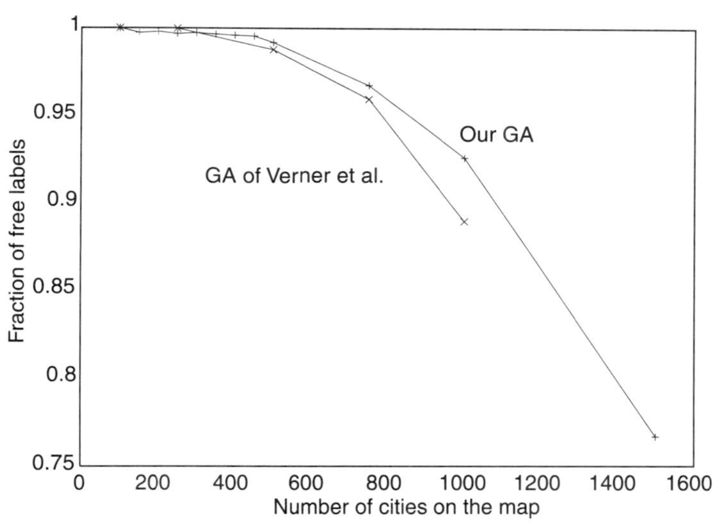

Figure 5-12 A comparison between our GA and the GA of Verner et al.

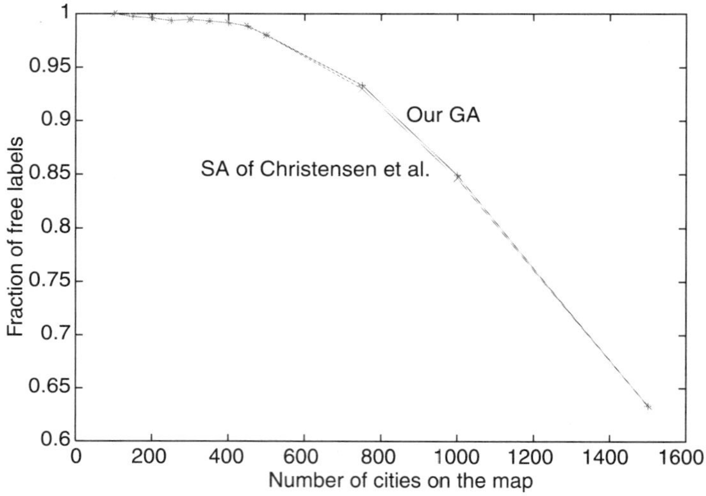

Figure 5-13 A comparison between our GA and the SA of Christensen et al.

- The simulated annealing algorithm of Christensen et al. produced solutions with equal quality to our GA. Also, for small maps our GA was faster but for large maps the SA was faster.

We also compared the SA against our GA when the problem became more difficult, for example when preferences for positions were considered or labels had to be selected for deletion. The SA produced lower quality solutions than the GA. However, the cooling schedule could be made longer (this would result in longer running times). No recommendations were made by Christensen et al. for more complicated problems.

Since the experimental study of van Dijk et al., at least four new algorithms have been proposed to solve the point-feature label-placement problem: another genetic algorithm developed by Raidl, an approach based on a formulation as a constraint satisfaction problem by Wagner & Wolf (1998), a branch-and-cut approach by Verweij & Aardal (1999), and a tabu-search approach by Yamamoto et al. (1999). They all seem to give good results, but we did not do an experimental comparison for these algorithms. We would like to stress, however, that in our opinion it is very important to have a way of extending the problem. Otherwise the algorithm will be of relatively little use in real-world applications where such soft constraints need to be considered. Adding more constraints to the problem can be easily handled by our GA through the use of geometrically local optimizers. Note, however, that they can be used because a GA is a population-based approach. It seems very difficult for the other algorithms we have seen to add other constraints.

Extensions of the GA for Map Labeling

In this section we will show how to extend the GA with additional constraints. Recall from the section on problems in GISs that the map-labeling problem

Table 5-2

Category	Size of Cities	Number in Data Set
Small	<50,000	1980
Medium	>590,000 and <1,000,000	394 – Washington
Mega	>1,000,000	6 + Washington

consists of maximizing the number of labels that do not intersect others. This is done by having as a fitness function a count of the number of free labels, and as geometrically local optimizer the slot-filling procedure. Now suppose we want to extend this problem with the following requirements:

1. Place the label in the preferred position, in the order of figure 5-4, whenever possible.
2. Make a distinction between normal cities and very important cities. The latter should always have a free label.
3. If the map is too crowded, it is possible to delete a label. This should be done automatically.
4. Labels of cities which have a high population size should not be deleted in favor of labels with a low population size.
5. The label of a city bordering the water should lie in the water.

To demonstrate the algorithm we used a data set of cities of the United States. The set contained 2380 cities, which we divided in three categories depending on the population size (table 5-2). A city with size smaller than 50,000 is called "small." A city with size between (and including) 50,000 and 1,000,000 is called "big." Cities with sizes beyond 1,000,000 are in the category "mega." The total number of cities in these categories is respectively 1980, 394, and 6. The cities in the mega category are always labeled (constraint 2). The city "Washington" is also guaranteed to be labeled, although it is a medium city. This is done by treating it as if it is a mega city.

We need to extend the geometrically local optimizer to include these constraints as described previously. As before, the city to which the geometrically local optimizer is applied has four slots, one for each position. Each slot can be in the following states:

- "empty" when no other label intersects it;
- "small" when the intersecting labels all belong to a small city;
- "medium" when the intersecting labels belong to at least one medium city and no mega city;
- "mega" when the intersecting labels belong to at least one mega city.

In order to satisfy constraint 4, cities are only allowed to place their label in a slot that is not occupied by the label of a larger city. Trying to fill slots is done in order of category. So, for example, a medium city tries to put its label in an empty slot. If none is available, it puts its label in a small slot (the small city will have to move its label or delete it later on in the run of the GA). If even that is not possible, the label is said to be "crowded." Handling of a crowded label is the

Figure 5-14 A labeling for cities in the United States. Light gray cities are small.

same for small and medium cities: they delete the label (this takes care of constraint 3). If the label of a mega city is crowded, we cannot delete it, since we have to satisfy constraint 2. A label of a mega city can only be crowded if another mega city completely covers it. In this case, nothing is done, since the covering label will be moved when the geometrically local optimizer is applied to the other mega city and the conflict will be resolved.

Constraint 1 is dealt with by trying to find a desired slot in the order of preference. If, for example, a medium city did not have empty slots and is now looking for a small slot, it does so in the order of preference. This leaves us with constraint 5. Unfortunately, the data set used did not have any information about bodies of water near the city. In order to demonstrate how this could be done, we declared the city of Los Angeles to have water in position 2 (top left) and 3 (bottom left). The order of preference was changed by giving positions that occupied water higher preference than positions that occupied land. This obviously only changed the order of preference for Los Angeles, which changed from 1, 2, 3, 4 to 2, 3, 1, 4.

The GA was run on a map consisting of cities of the United States, and the resulting labeling is depicted in figure 5-14.

Discussion

When designing algorithms to solve hard problems arising in the use of a GIS, it is necessary to take the characteristics of these problems into account. Important are the following points:

- The problems have to find solutions in a very large search space. A powerful combinatorial solver is therefore needed, such as a genetic algorithm, simulated annealing, tabu-search, and so forth.
- Geographic problems often have to incorporate other, soft constraints as well. Just solving the combinatorial part of the problem might be of relatively little use if it cannot extended with other constraints.

- The structure of the problem is often quite clear. This is because the dependencies between the problem variables are geometrically determined. For a GA this translates to insight in the linkage of building blocks.
- Users of a GIS will need to run the algorithm and do not want to deal with parameters that are a result of the algorithm used.

Keeping the above points in mind, we can try to compare algorithms. Advantages of the approach described in this chapter are:

- Extending the GA with other constraints is relatively easy by using the geometrically local optimizers.
- The linkage of the problem was exploited by using a cross-over operator that mixes on the level of the building blocks, and also by the geometrically local optimizer, which explicitly tries to create building blocks.
- No tuning of weighting parameters in the fitness function is required.
- Except for the population size, the algorithm itself has no parameters that need to be set.

However, the resulting GA has the disadvantage that it can be time consuming. In part this holds for all combinatorial solvers, because they are trying to find close to optimal solutions. Parallelization can be a solution, since it is very easy to parallelize GAs. There is also an aspect of the GA which can be considered both an advantage as a disadvantage: the need (or possibility) for population sizing.

The GA we designed to solve the map labeling problem has one parameter that has to be set: the population size. Enlarging the population size improves the quality of the solutions, but will cause longer running times. Also, beyond a certain value enlarging the population size will not cause any improvement. One can try to set the population size at the optimal population size by using various methods:

- *Reckoning*: try to set the population size by trial and error. If enlarging the population size gives an improvement, continue increasing it. After a while the user of the GA gets a feeling for which values work best.
- *Analysis*: try to develop a model that captures the relation between the input and the required population. See, for example, Goldberg et al. (1992) and Harik et al. (1997).
- *Adaptive resizing*: automate the process of reckoning and let differently sized GAs compete. See, for example, Harik & Lobo (1999) and Smith (1993).

Analysis is mostly done in a theoretical setting and is difficult to do for real-world applications. Adaptive resizing is nice in that it does not require the user to set the population size but will consume up additional computation time to find the optimal population size. Reckoning is probably the method most used, but it will probably give a population size that is a tradeoff between time used and quality produced.

Which method to use depends on the user. Some users will find having to set a parameter (resulting from the algorithm used) a bother. In that case, this last parameter can be eliminated by using adaptive resizing, but it will require a certain cost. On the other hand, if the parameter is kept, the user can have direct control on the amount of computational resources which are used (and thus the running time). This can be an advantage, since it is possible to make a tradeoff between the amount of time available and the solution quality required. What is best depends on the situation.

Conclusion

When designing genetic algorithms that have to search in large search spaces, it is important to understand what makes the problem hard. This chapter stressed the role of *linkage* between genes, which is a result of nonlinear dependencies between problem variables. It was explained why linkage makes a problem hard and how a problem can be seen as composed of building blocks: groups of genes with strong linkage between them. Problems that arise when working with geographic information systems often can be solved with a GA because the linkage is geometrically determined. This knowledge about the linkage of the building blocks can be exploited and used to design an efficient GA. This is done by designing a crossover that mixes on the level of building blocks and by adding a novel operator (the geometrically local optimizer) that explicitly creates good candidates that could be building blocks. However, besides making the design problem easier by offering extra information about the GIS problem, the fact that we are dealing with a GIS makes the design problem also harder. This is because users want a robust GA which is extendible, has few parameters and allows for expressing domain knowledge. These demands are met by using the geometrically local optimizer, an operator that works on a geometrically local scale, and allows for an easy handling of local constraints without the need to alter the fitness function.

The expression of the ideas in this chapter resulted in a GA for a specific problem: labeling a map consisting of point features. This GA compared favorably against other algorithms for map labeling, and it was shown how to extend the GA to include other constraints.

Notes

Steven van Dijk is supported by the Dutch Organization for Scientific Research (NWO).

1. The geometrically local optimizer should not be confused with "local search" (such as a hillclimber). The difference is that local search searches locally with respect to the fitness landscape whereas the geometrically local optimizer searches locally with respect to the map. The combined appliances of geometrically local optimizers can be seen as a local search.

References

Christensen, J., J. Marks, & S. Shieber. 1995. An empirical study of algorithms for point-feature label placement. ACM Transactions on Graphics, 14(3), 203–232.

Goldberg, D., K. Deb, & J. Clark. 1992. Genetic algorithms, noise, and the sizing of populations. Complex Systems, 6, 333–362

Harik, G. & F. Lobo. 1999. A parameter-less genetic algorithm. In: W. Banzhaf, J. Daida, A. Eiben, M. Garzon, V. Honavar, M. Jakiela, & R. Smith (eds.), GECCO-99:

Proceedings of the Genetic and Evolutionary Computation Conference, July. Morgan Kaufmann.

Harik, G. E. Cantú-Paz, D. Goldberg, & B. Miller. 1997. The gambler's ruin problem, genetic algorithms, and the sizing of populations. In: Proceedings of the IEEE International Conference on Evolutionary Computation, 7–12.

Hirsch, S. 1982. An algorithm for automatic name placement around point data. The American Cartographer, 9(1), 5–17.

Kargupta, H. 1996. The gene expression messy genetic algorithm. In: Proceedings of the IEEE International Conference On Evolutionary Computation. Nagoya, Japan.

Marks, J. & S. Shieber. 1991. The computational complexity of cartographic label placement. Technical Report TR-05-91, Harvard CS.

Smith, R. 1993. Adaptively resizing populations: An algorithm and analysis. Technical report, University of Alabama, February.

Thierens, D. & D. Goldberg. Elitist recombination: An integrated selection recombination GA. In: Proceedings IEEE International Conference on Evolutionary Computation, 508–512. IEEE Service Center, Piscataway, N.J.

van Dijk, S., D. Thierens, & M. de Berg. 1998. Robust genetic algorithms for high quality map labeling. Technical Report TR-1998-41, Utrecht University.

van Dijk, S., D. Thierens, & M. de Berg. 1999. On the design of genetic algorithms for geographical applications. In: W. Banzhaf, J. Daida, A. Eiben, M. Garzon, V. Honavar, M. Jakiela, & R. Smith (eds.) GECCO-99: Proceedings of the Genetic and Evolutionary Computation Conference, July, Morgan Kaufmann.

Verner, O., R. Wainwright, & D. Schoenefeld. 1996. Placing text labels on maps and diagrams using genetic algorithms with masking. INFORMS Journal of Computing, 9(3), 261–275.

Verweij, A. & K. Aardal. 1999. An optimisation algorithm for maximum independent set with applications in map labelling. In: Procedings 7th Annual European Symposium on Algorithms.

Wagner, F. & A. Wolf. 1998. A combinatorial framework for map labeling. In S.H. Whitesides (ed.), Proceedings of the Symposium on Graph Drawing (GD '98), 13–15 August. Lecture Notes in Computer Science, Vol. 1547, 316–331. Springer-Verlag.

Yamamoto, M., L. Lorena, & G. Camâara. 1999. Tabu search application for point features cartographic label placement problems. In Proceedings 3rd Metaheuristics International Conference (MIC'99), Angra dos Reis (to appear).

Zoraster, S. 1986. Integer programming applied to the map label placement problem. Cartographica, 23(3), 16–27.

6

Evolutionary Modeling of Routes

The Case of Road Design

ÂNGELA GUIMARÃES PEREIRA

In this study a route is defined as the path that a linear structure or facility follows in the terrain. Linear structures comprise facilities such as roads, motorways, railways, pipelines, electrical power lines, and telephone cables, each of these structures requiring specific technical parameters in what concerns the geometry of the path and having different effects on the terrain they traverse. Amongst these structures, roads and motorways are the group that creates the greatest overall impact; accordingly Portuguese legislation requires an environmental impact assessment (EIA) process as part of the necessary licensing approval. Usually the alternative (or alternatives) that undergo the EIA process is justified in terms of technical and economical issues. The result is that if major environmental impacts are identified by the EIA study, a myriad of mitigation measures are proposed, very seldom the redesign of the path being carried out (Guimarães Pereira & Antunes, 1996). Preliminary studies that precede the implementation of these types of projects are technically detailed and often come together with economical feasibility studies, shelving environmental issues for later assessment. In the methodology proposed in this chapter a multidimensional evaluation methodology, *multicriteria evaluation,* will be combined with the robustness of a search methodology, *genetic algorithms* (GAs) to generate alternative road routes that take into consideration environmental, economical, technical, and social criteria. These criteria are referenced to the physical space where the road is to be placed and therefore this methodology is embedded into a geographic information system (GIS).

Genetic algorithms are particularly attractive to apply to multi-modal problems, allowing the exploration of spatial features to eventually find "best compromise" alternatives because these algorithms proceed their search by maintaining a population of solutions, that they can simultaneously exploit for their efficiency.[1] Moreover, the particular *mixing* mechanism provides the means to recombine solutions and explore the search space.

The remainder of this chapter describes evolutionary modeling of road routes, in particular the coding onto a GA of the geometric algorithm that accounts for the technical aspects of motorway siting. The details of the implementation of the MCDA-GA methodology, running within the GIS GRASS 4.1 (Geographic Resources Analysis Support System) and its application to generate and evaluate alternative routes of a section of a Portuguese complementary itinerary (IC7) will be presented.

Route Geometry of Roads

Levels of Evaluation of Routes

Routing algorithms available in the GIS toolbox fall mainly into a class of standard algorithms for finding optimal paths on terrain with a known cost map, namely wavefront propagation algorithms. A review of shortest path algorithms has been done elsewhere (Guimarães Pereira, 1996, 1997).

In practice, the most usual way of planning routes is carried out in two phases: routes are sketched on a map, normally on a 1 : 25,000 scale, and then a topographer goes into the field to check how feasible the routes are, making the necessary adjustments. A route is designed according to many technical parameters and political guidelines. However, in this design only a few environmental and social constraints are considered—these are the object of other assessment studies (e.g., EIA). Seldom do these procedures lead to more than one candidate route. Indeed, there is no comprehensive method to generate alternative routes that take into account several aspects of the problem. Probably, that is due to the fact that there are no adequate computational tools to carry out the job in such a comprehensive way.

Some GISs provide *wavefront-propagation-based* tools to determine shortest paths through a *surface* of cumulative cost cells or any derived land suitability raster map (a result of reclassification of land information into a suitability surface by mathematical, logical, or other combination of mapped information). Because of the many passes involved in such a process these tools are extremely time consuming. Many inadequacies have also been identified concerning the use of such algorithms in suitability surfaces (Rowe & Richbourg, 1990; Janssen & Rietveld, 1990).

One of the most important drawbacks of these algorithms is that they proceed sequentially and so the geometry of the path cannot be evaluated as a whole. Therefore criteria like homogeneity and smoothness of the path, overall cost of the path, and, in general, all criteria that evaluate the path as a single object, cannot be evaluated until the complete path is established. In a way it can be argued that these algorithms do not directly integrate a technically compliant geometry of the path.

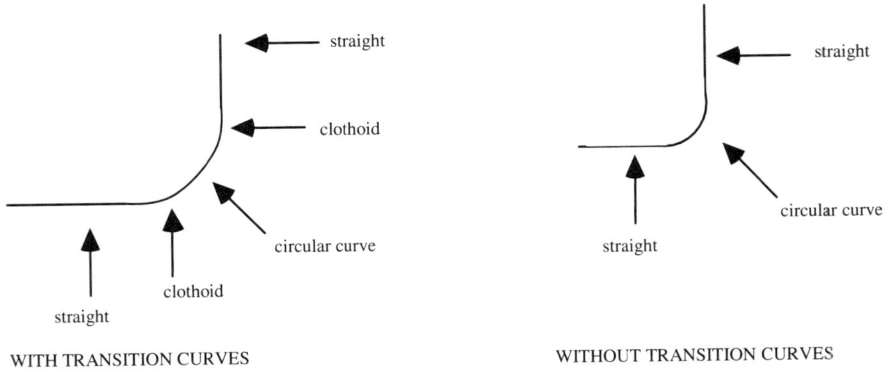

Figure 6-1 Geometric elements of a road's plan view for the cases when there are transition curves and when there are not.

The use of GAs can be of great help to largely overcome this drawback. Indeed, three levels of evaluation can be performed for each route and aggregated with multicriteria evaluation methods:

1. as a single object;
2. for its geometric elements (on the plan view and longitudinal profile);
3. for the cells it traverses.

It is illustrated below ("Road Geometry") how the codification of the routes is itself the geometric representation of the path: the representation of the route as a single object allows the evaluation of it as a whole. The path is a sequence of geometric elements that can be evaluated with appropriate technical criteria. Finally, the representation of the routes by its traversal cells allows for the evaluation of the path in the terrain it occupies.

In the remainder of this section the *geometric* algorithm developed to site motorways and, in general, roads whose design speed is greater or equal than 120 km/h, will be described.

Road Geometry

Although roads are designed according to the cross-section, longitudinal section, and plan view, here only the last two are considered, for the former adds sophistication that the proposed generation of alternatives method does not intend to achieve. In other words, the cross-section involves too detailed a design, which is not the aim of this methodology, as pointed out earlier.

Plan View

Basically roads are considered as tuples of geometric elements, as depicted in figure 6-1. In the plan view there are two basic elements, a rectilinear element and a circular element. The rectilinear element, called the *straight*, is a line for which a range of suitable extensions is set according to the design speed of the

Table 6-1 Geometric elements and parameters of the path

Design Speed km/h	Straight		Circular Curves	
	Maximum Extension (m)	Minimum Extension (m)	Minimum Curvature Radius (m)	Minimum Curvature Extension (m)
120	2400	720	1000	250
130	2600	780	1200	320
140	2800	840	1400	400

Adapted from Portuguese Standards NP3-91 (JAE, 1992).

road. The circular element comprises either one or three subelements: two transition curves (an "anterior" clothoid and a "posterior" clothoid) and a circular curve. The existence of the transition curves—designed as clothoids—depends upon the curvature radius of the circular curve.

Table 6-1 summarizes the suitable values for the technical parameters of a road's plan view, for three different design speed values.

Longitudinal Section

In the longitudinal section there are also three basic elements: the gradient, vertical crest curves, and vertical sag curves. The definition of this profile takes into account the topography of the terrain, the plan view, and safety parameters, such as visibility. The interest of this profile is the coordination between the slope of the elements and the form of the elements given by the plan view. There is a compromise to attain between the slope of the straight and its extension. The Portuguese standards NP3-91 (JAE, 1992) establish critical extensions for the straight.

There are two types of vertical curves: the vertical crest curves and the vertical sag curves. Similar to the gradient case, there has to be a good coordination of those curves and the plan view. Figure 6-2 illustrates the elements of the longitudinal section. In table 6-2, values of the parameter values that characterize the elements of the longitudinal section are summarized for two design speeds.

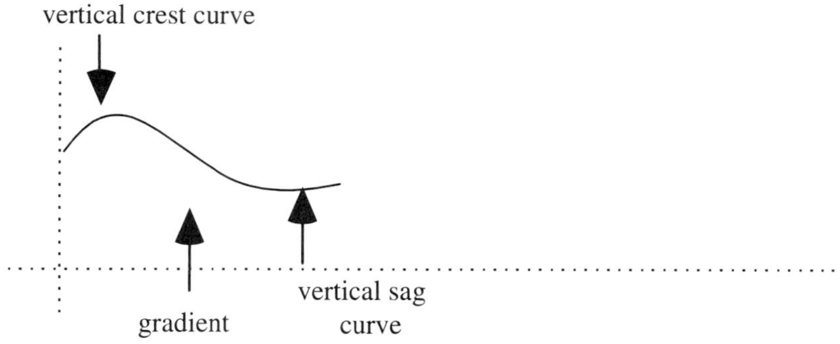

Figure 6-2 Geometric elements of the longitudinal section.

Table 6-2 Fundamental parameters of the longitudinal section

	Gradient		Vertical Curves—Minimum Radius	
Design Speed (km/h)	Slope (%)	Critical Extension (m)	Crest (m)	Sag (m)
120	4	300	16,000	7000
140	3	420	20,000	8000

Source: JAE (1992).

Modeling Route Geometry into Genetic Algorithms' Data Structure

Codification of Routes

In this study, a route is addressed as a concatenation of points, since it must be evaluated at all traversal points according to the chosen spatial criteria mapped into raster layers. In this way the loss of information can be minimized compared with procedures where it has to be aggregated in order to reduce complexity.

However, the codification of such a chain as a concatenation of single coordinates would lead to computational inefficiency. Therefore another type of codification is implemented, based upon the geometric components of the route's plan view.

Hence, to model routes by means of GA's data structures, each route is imagined as a sequence of the plan view's geometric elements (linear and curvilinear), each characterized for its parameters, as many times as is necessary to reach the length between the target points. Figure 6-3 illustrates the structure of the routes codified into bit strings for the GA.

Decodification of Routes

Figure 6-4 illustrates the decodification process of road routes. For instance, the route length of a simple two-tuple section of a high-speed type of road (design speed is 120 km/h) designed on a 100 m resolution map is approximately 6 km. For the chosen design speed each individual route is codified into a 46-bit long binary string. The decodification of the string is made in three steps:

1. The geometric parameters are decodified, their values and the coherence among them being checked against the feasible ranges of the technical criteria

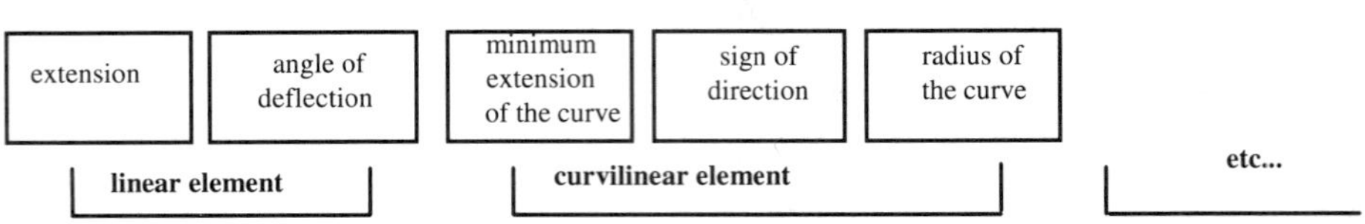

Figure 6-3 Codification of the route parameters using GAs.

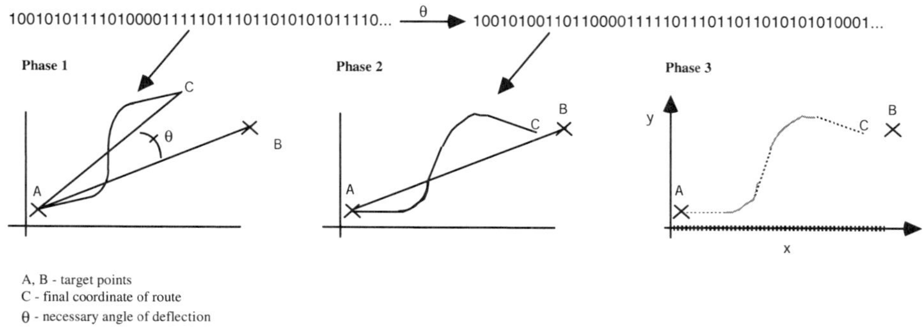

Figure 6-4 Steps of decodification of routes in the plan.

previously set. From these calculations the route's ending coordinates are computed.

2. The route is adjusted for the axis that links the target coordinates; this is carried out by spatial rotation of the route, which consists of the adjustment of the angles of deflection of the geometric elements by some value θ. This modification is immediately transposed into the binary coding by modifying the codified angles of deflection.
3. The route traversal cells are derived, with reference to the initial target coordinate.

The resolution of the georeferenced information will determine the degree of detail with which the route is represented.

The decodified route consists of a sequence of the plan view's geometric elements: a straight followed by a transition curve (clothoid), a circular curve and another transition curve (clothoid). The length and therefore the number of traversal cells of the candidate path may vary, that is, the specification of routes will vary according to the path's length, which is determined by the values of the extension parameters of both the straight and the curve. The *plotting* of these elements on the terrain gives the set of cells that the route traverses. In this way the route can be assessed for its technical characteristics, as well as for its *georeferenced* characteristics. The details of the decodification of routes in the plan were exemplified in the earlier in this section.

The number of bits in the codification of each of the parameters depends essentially upon the resolution of the raster maps used in the evaluation of the cells traversed by the route. Indeed, the degree of detail of the route representation in the plan is dependent upon the resolution of the raster data. Table 6-3 illustrates how the binary string size is related to the resolution of the raster data indicating the number of bits to codify the road plan view parameters when the road's design speed is 120 km/h.

Of course this only applies to the genes that codify distances, that is, extensions and radius parameters, since the angle of deflection of the straight is independent of the resolution of the raster data, 2π. When considering 6 bits to codify the angle of deflection only 0.1 rad of resolution to this parameter is actually allowed.

Table 6-3 How the number of bits in the string relates to the resolution of the maps. These values refer to a design speed equal to 120 km/h

Parameter		Resolution: 100 m		Resolution: 50 m	
Name	Maximum Value (m)	Multiplying Factor	Number of Bits	Multiplying Factor	Number of Bits
Straight extension	2400	100	5	50	6
Curve radius[a]	>1000	1000	6	50	7
Curve[b] extension	>250	100	5	50	6

[a]Curve radius can be more than 5000 m, in which case transition curves are not needed. It was assumed that this parameter may take values not much higher than this.
[b]It was assumed here that the extension of the curve is not longer than the straight extension, which is reasonable according to the Portuguese Standards.

Representation of the Geometric Elements in the Cartesian Coordinate Plan

Straight Representation

The straight representation in the plan is straightforward. Taking the initial coordinates (x_i, y_i) and the angle of deflection, the final coordinates (x_S, y_S) are calculated with simple parametric line equations (see, e.g., Bowyer & Woodwark, 1983):

$$\begin{aligned} x_S &= x_i + (E.S.)\cos\alpha \\ y_S &= y_i + (E.S.)\sin\alpha \end{aligned} \tag{1}$$

where $E.S.$ is the extension of the straight and α the angle of deflection of the straight.

Intermediate coordinates are calculated by incrementing the distance by a factor equal to the raster resolution. The last coordinate, (x_S, y_S) is the initial coordinate of the next geometric parameter, one of the three following cases being possible:

1. Clothoid preceding circular curve;
2. Circular curve if curvature radius is higher than 5000 m;
3. Another straight.

Transition Curve Representation

As noted earlier, the transition curves are clothoids (Kanayama & Miyake, 1986; Nelson, 1989). If case 1 holds (see above), the clothoid parameter A is estimated as described in JAE (1992) and chosen as explained there. From this estimation, the final coordinates of the anterior clothoid can be calculated:

$$\begin{aligned} x_{aC} &= x_S + L\cos\beta - \mathscr{P}\sin\beta \\ y_{aC} &= y_S + L\sin\beta + \mathscr{P}\cos\beta \end{aligned} \tag{2}$$

where $L = A^2/R$ is the length of the clothoid, as indicated in the Portuguese Standards (JAE, 1992), R is the radius of the circular curve, β is the angle of deflection of the geometric elements that precede the transition curve, and finally, $\mathcal{P}$ is the sense of deflection of the circular curve.

As in the case of the straight, these coordinates (x_{aC}, y_{aC}) are the initial coordinates of the next geometric parameter, that is, the circular curve. Intermediate coordinates are estimated by incrementing the distance by a factor equal to the raster resolution.

As suggested in the Portuguese Standards (JAE, 1992) the value of A is set to be equal for both anterior and posterior clothoids, as well as the length L. The posterior clothoid can only be represented in the plan after the circular curve that precedes it is itself represented. The procedure to estimate the ending coordinates (x_{pC}, y_{pC}) of this geometric element is, however, similar to that for the anterior clothoid:

$$\begin{aligned} x_{pC} &= x_{CC} + L\cos\gamma \\ y_{pC} &= y_{CC} + \frac{L^2}{6.0R}\sin\gamma \end{aligned} \tag{3}$$

where (x_{CC}, y_{CC}) are the ending coordinates of the circular curve that precedes the clothoid and γ is the total deflection of the elements that precede the clothoid.

Anterior and posterior clothoids preferably have different senses of deflection, the anterior following the sense of the circular curve and the posterior following the opposite sense.

Circular Curve

The first coordinates of the circular curve are the last coordinates of the element that precedes it, either the anterior clothoid (case 1), (x_{aC}, y_{aC}) or the straight (case 2), (x_s, y_s). The extension and the curvature radius together with the sign of deflection, as codified in the binary string, completely define the trajectory of the circular element. First the origin of the circumference (O_x, O_y) is calculated, which depends upon the geometric element that precedes the curve:

1. Straight preceding the curve:

$$\begin{aligned} O_x &= x_i + R\cos\varphi \\ O_y &= y_i + R\sin\varphi \end{aligned} \tag{4}$$

where (x_i, y_i) is the origin of the route and φ is the composed angle that represents the total deflection of the preceding elements and the curve deflection angle itself.[2]

2. Clothoid preceding the curve:

$$O_x = x_i + \frac{L}{2}\cos\varphi - \mathscr{P}\left(\frac{L^2}{24R} + R\right)\sin\varphi$$
$$O_y = y_i + \frac{L}{2}\sin\varphi + \mathscr{P}\left(\frac{L^2}{24R} + R\right)\cos\varphi \tag{5}$$

where symbols have the meaning described earlier.

The final coordinates can be computed using the parametric equations of a circle:

$$x_{CC} = O_x + R\cos\upsilon$$
$$y_{CC} = O_y + R\sin\upsilon \tag{6}$$

where υ is the composed angle of deflection of the curve including the codified angle of deflection and that of the total deflection of the preceding geometric elements. The intermediate coordinates of the curve can be calculated by incrementing the distance by a factor equal to the raster cells resolution.

As before, the last coordinates, (x_{CC}, y_{CC}) will be the first coordinates of the next geometric element, a clothoid, or a straight, depending on the curve's circular radius.

Multicriteria Evaluation Genetic Algorithm

Multiobjective and Multicriteria Evaluation

Multiobjective optimization of vector-valued functions through genetic algorithms has been studied by some authors (Wienke et al., 1992; Fonseca & Fleming, 1993; Horn & Nafpliotis, 1993). In particular, multiattribute utility analysis (MAUA) (Keeney & Raiffa, 1976; Keeney, 1992) was used to achieve the set of points known as the Pareto-optimal set.

The GA-MAUA implementations were developed for the multiobjective model, that is, for the continuous case of the multicriteria models. The siting problems addressed here cannot be modeled as continuous problems because the search space is finite. Therefore a new methodology was developed that applies to discrete multicriteria problems.

Evaluation Methodology

Genetic algorithms are a versatile type of algorithm that may accommodate evaluation functions (fitness function) of any type because of their general purpose nature.

The evaluation methodology presented here has some analogies with the $P\gamma$ prescription of multicriteria decision analysis (MCDA) problems (Roy & Vincke, 1981; Vanderpooten, 1990). Evaluations are made in terms of *suitability to carry out an action.*

Preliminary Definitions

The formal basis and useful definitions for the description of the $P\gamma$ strategies pointed out above are described in this section. With some modifications the same model as in decision aid methods, *Alternatives, Attributes, Evaluators* (Vansnick, 1990) is used here:

1. The set of alternatives[3] is finite, though large, $A = \{a_0, \ldots, a_N, \ldots, a_S\}$, $S \to \infty$, too unmanageable to perform comparisons on a matrix calculation basis. The current set of alternatives as a subset of A, $P = \{a_0, \ldots a_N\}$, of size N with $P \subset A$.
2. Set attributes $\{X_1, \ldots X_\varphi\}$ of size φ; each attribute is a set of at least two levels of some underlying dimension.
3. Set evaluators $\{\Gamma_1, \ldots, \Gamma_\varphi\}$ of size φ; an evaluator $\Gamma_i: A \to \delta(x_i) - \phi$ is a function that associates with each alternative an evaluation measure for each dimension i.

Condition 1 is the motivation to integrate genetic algorithms with the multicriteria model.

Evaluation Procedure

In the ranking procedure, comparison among alternatives pertaining to the current population is carried out in order to obtain a rank. This implies *pairwise comparison* of each alternative with the whole set of alternatives for the whole set of criteria. Each action is assigned a value of strength for the problem's context, inspired by the PROMETHEE methods (Brans et al., 1986; Brans & Mareschal, 1990), a subfamily of the outranking methods. Eventually alternatives are ranked and assigned a value of relative strength within the set.

This formulation is useful when potential actions have to be differentiated according to their relative interest, that is, their goodness for the action's objectives. Integration of multicriteria concepts with GAs has the most usefulness in this procedure.

When adopting this strategy for the generation of alternatives process, it implies establishing at each GA generation preference relations within the set of alternatives. Hence, this means establishing the *strength*[4]—which is a relative measure—of each candidate to be used in the GA selection procedure.

In practice, alternatives are compared in pairwise comparison fashion for each of the criteria with all N alternatives in the current population P. The comparison consists of computing the difference between the *intensity of preference* on each criterion for each pair of alternatives a and a', $a, a' \in P = \{a_1, \ldots, a_N\}$. The aggregation of these differences is computed through a simple sum, resulting in $2(N-1)$ measures, each of which represents how alternative a is preferred to the other alternative it is being compared to (a positive measure) and how alternative a is not preferred to the other alternative it is being compared to (a negative measure). Through the sum of these values into a positive overall measure and a negative one, the strength of alternative a within the population can be computed.

Alternatives are ranked in terms of preferences defined over the criteria scores. As mentioned earlier, criteria scores represent *suitability values to carry out the action*... according to a certain point of view. Suitability values for each of the criteria are considered as ranging from 0 to 1. If the alternative scores 0 for a specific criterion, it means that it is not suitable[5] according to that criterion, whereas if the alternative scores 1, it means that the alternative is suitable—thus it is a maximization procedure. This assignment is not necessarily binary, all the intermediate values being possible in a continuous or discrete scale of suitability for the criterion. These values are actually read as membership values to a fuzzy set "suitable to site" because each criterion is assigned a membership function that may have different forms.

Formal Definition

Let $\Gamma_i^j\; i = 1, \ldots, \varphi$ and $j = 1, \ldots, N$ be the ith criterion score, with $\Gamma_i^j \in [0, 1]$, for the jth alternative. Each alternative is compared with all the other ones in the set P, for each criterion to give a measure of preference difference:

$$\Gamma_1^a - \Gamma_1^i, \text{for } i = 1, \ldots, a, \ldots, N \text{ for criterion } 1$$

$$\Gamma_2^a - \Gamma_1^i, \text{for } i = 1, \ldots, a, \ldots, N \text{ for criterion } 2$$

$$\vdots$$

$$\Gamma_j^a - \Gamma_j^i, \text{for } i = 1, \ldots, a, \ldots, N \text{ for criterion } j, \quad \therefore\; j = 1, \ldots, \varphi$$

These pairwise comparisons hold the following properties:

1. If $\Gamma_j^a - \Gamma_j^{a'} = \varepsilon$, with $\varepsilon \in [-E, E]$, a is indifferent to a' for the jth criterion, if an indifference threshold E is applied.
2. If $\Gamma_j^a - \Gamma_j^{a'} > \xi$, with $\xi \geq E$, a is preferred to a' for the jth criterion and therefore a is stronger than a' for that criterion, if a preference threshold ξ is applied.
3. If $\Gamma_j^a - \Gamma_j^{a'} < \xi$ with $|\xi| \geq E$, a is preferred by a' for the jth criterion and therefore a' is stronger than a for that criterion, if a preference threshold ξ is applied.

For simplicity, just the case where E and ξ are zero will be considered.

These differences are simply summed for each criterion. Thus for each pair of alternatives, a and a', two measures are computed from the preference differences between the two alternatives:

$$\mu_{a'}^{+}(a) = \sum_{j=1}^{\varphi^{+}} (\Gamma_j^a - \Gamma_j^{a'}) \text{ if } \Gamma_j^a - \Gamma_j^{a'} > \xi \tag{7}$$

where φ^{+} is the number of criteria for which $\Gamma_j^a - \Gamma^{a'} > \xi$ and

$$\mu_{a'}^{-}(a) \sum_{j=1}^{\varphi^{-}} (|\Gamma_j^a - \Gamma_j^{a'}|) \text{ if } \Gamma_j^a - \Gamma_j^{a'} < |\xi| \tag{8}$$

where φ^- is the number of criteria for which $\Gamma_j^a - \Gamma_j^{a'} < \xi$. When $\Gamma_j^a - \Gamma_j^{a'} = \varepsilon$, it means that both alternatives score the same preference for the jth criterion and therefore that specific criterion does not influence the dominance relation between both alternatives.

Hence, there will be $(N-1)$ measures $\mu^+(a)$ and $(N-1)$ measures $\mu^-(a)$ for each alternative a, that describe respectively how good and how bad alternative a is compared with each of the others in the set.

The strength Φ of the alternative in the current set is defined as the difference between the positive strength Φ^+ and the negative strength Φ^- for all criteria and all alternatives. The Φ^+ is the measure of how stronger alternative i is for the criteria where it scores higher than the other alternatives to which it is compared:

$$\Phi^+(i) = \frac{1}{N-1} \sum_{j=1}^{N} \mu_j^+(i) \tag{9}$$

and Φ^+ is the measure of how much weaker alternative i is for the criteria where it scores lower than the other alternatives to which it is compared:

$$\Phi^-(i) = \frac{1}{N-1} \sum_{j=1}^{N} \mu_j^-(i) \tag{10}$$

$\mu_j^+(i)$ and $\mu_j^-(i)$ result from pairwise comparison of i and all the other alternatives in the set for all criteria as described in equations 9 and 10 respectively.

To aggregate these measures $\Phi^+(a)$ (equation 9) and $\Phi^-(a)$ (equation 10), respectively the positive and negative strengths of alternative a within the set P are calculated and finally the overall strength measure $\Phi(a)$ is achieved by calculating the difference between those two:

$$\Phi(a) = \frac{\Phi^+(a) - \Phi^-(a)}{N} \tag{11}$$

To report these values onto the interval [0, 1], the necessary scaling adjustments are done.

Algorithm Implementation

Below is presented pseudocode for the multicriteria evaluation fitness function in the GA, for this strategy. The number of alternatives is the population size.

```
fitness(population)
{  for (alt=number of alternatives){
     for (oth=number of alternatives){
       for (nr=number of criteria){
         Pref_difference=Pref_nr(alt) - Pref_nr(oth);
         if (Pref_difference>0) {
           miu+_alt,oth+=Pref_difference;
```

```
            count_pos_crit ++;}
          if (Pref_difference<0){
            miu-alt,oth+=|Pref_difference|;
            count_neg_crit++;}
            }
            Fi+alt+=miu+alt,oth,
            Fi-alt+=miu-alt,oth,}
          Fialt=1/N(Fi+alt - Fi-alt);
        }
  }
```

Generation of Alternative Routes for a Section of the IC7 in the Region of Covilhã

Generalities

The IC7 is a complementary itinerary that will be about 59.425 km long and will link Venda de Galizes to Covilhã, a town located in the center of Portugal. From the five sections into which the IC7 was divided, the route section studied here is approximately 10 km long (linking Venda de Galizes to Vide). The IC7 traverses a region that is orographically difficult and extensively forested. Although it is not exactly a motorway, the design speed is 120 km/h and therefore the technical and safety requisites fall into the category of the roads of that design speed.

This road aims at providing a transversal route into the central region of Portugal, in order to revitalize its development. According to the "nontechnical" summary of the environmental impact assessment study for the IC7, there are no significant negative impacts in what concerns land-use planning, the major impacts being crossing of areas with high agricultural aptitude and of small villages and also at "scenery" level because of visual impacts. At faunistic level, the fractionating of habitats constitutes a major negative impact.

Structuring Issues

The structuring issues considered here are the choice of criteria upon which the alternative routes may be generated and evaluated, as well as the construction of such criteria (i.e., building criteria scores). Criteria are built according to the points of view involved in the construction of a road of this type. Therefore, in the following sections, actors and points of view are identified, followed by the actual choice of the criteria that will sustain the search exercise that will then be carried out.

Actors and Points of View

Five main points of view may be identified which pertain to different actors. The mainstream concerns come from the following points of view:

1. *Technical*: engineering, design, and safety.
2. *Environmental*: impacts and risk to the environment.
3. *Social*: impacts such as nuisance, noise, and land ownership.
4. *Economic*: aspects such as lobbies, infrastructure and transportation costs, land devaluation, and regional and area accessibility.
5. *Political*: lobbies, policies, voting, and so forth.

The actors associated with these points of view are myriad. Among them can be found road authority representatives; national government; design and project engineering; economists; environmental experts and groups; land owners; communities of the surrounding areas; political groups; and regional and central government.

Hence, there are a large number of actors in this process. For the purposes of generating alternatives, of these points of view, numbers 1–4 are included as basic starting points to evaluate possible routes, since they provide the means to credibly justify alternative routes for further thorough analysis.

Table 6-4 is a simplified list of the criteria applied to evaluate the routes. These are divided according to the dimension of analysis they represent. Because of data availability the analysis made here may not be so wide, and it must be noted that the evaluation of the routes is strongly biased towards its technical aspects, since the data required are available.

All georeferenced data sets are stored at $0.0625\,\text{km}^2$ resolution in a $0.25 \times 350 \times 200\,\text{km}^2$ grid in the GIS GRASS (CERL, 1993). All data layers were created after official data on paper maps (1:25,000 scale *Cartas Militares* provided by the Serviço Cartogáfico do Exército) and also after satellite images.

Construction of Criteria

Three main groups of criteria are addressed here to deal with motorway siting (see table 6-4). The first one regards the technical dimension of the problem, which includes mainly design and safety criteria; the second regards economic aspects, and the third concerns environmental issues. The first two groups are usually contained at very early stages of the conception of the project, while the third is often totally disregarded at these stages.

In the conceptual design of this system to generate alternatives, not only technical and economic aspects but also environmental aspects were taken into account, mostly georeferenced criteria. With the aid of a GIS the whole path can be evaluated for the terrain it traverses. Figure 6-5 illustrates the georeferenced criteria used in this exercise. This is the main contribution of a GIS-based search engine to generate alternative routes. Criteria pertaining to other actors can be added, this system being especially concerned with those that can be georeferenced.

Technical criteria are regulated by national standards. Portuguese standards for building motorways and main roads are the Norma Portuguesa P2-91 and P3-91 (JAE, 1992), which regulate the technical aspects of a route with regard to its geometry and safety. It is on the basis of these standards, which already incorporate safety and technical elements, that routes are codified and partially evaluated for their technical feasibility.

Table 6-4 Criteria scores used to generate and evaluate alternative routes for the IC7 section between Venda de Galizes and Vide

Point of View	Criterion	Description	Scores	Data Requirements
Technical	Length	(1) Length $\leq 2 \times \sqrt{(X_j - X_i)^2 + (Y_j - Y_i)^2}$ (2) $(X_F, Y_F) = (X_{FT}, Y_{FT})$	(1) 0.0 if $l >$ length 1.0 if $l \leq$ length (2) Preference is 1.0 if coordinates are coincident and decays linearly with distance	Decodification of the route into its linear and circular elements
	Smoothness	Sequential circular curves run in opposite direction	1.0 if opposite direction 0.5 if same direction	Decodification of the route into its circular elements
	Safety	All curve parameters within feasible ranges	Inside the ranges 1.0; outside the ranges linear decaying function (flat membership function); plan view for VB = 120 km/h: Circular Curve ≥ 250	Decodification of the route into its circular parameters
	Homogeneity	All straight parameters within feasible ranges	Inside the ranges 1.0; outside the ranges, linear decaying function; plan view for VB = 120 km/h: $270 \leq$ Straight ≤ 2400	Decodification of the route into its linear parameters
	Coordination plan view and longitudinal section for straights	Plan view parameters and longitudinal section parameter coordination within the useful range, for straights	If concordant, 1.0; if not concordant, linear decaying function of the parameter values	Slope and height raster image sand decodification of the route
	Slope	Slope has influence in many of the criteria above described	1.0 if inside range $\frac{1.0}{\text{slope (\%)} \times 100}$ otherwise	Slope raster image
Environmental	Vegetation cover and land use	Endemic forest and agricultural land	0.0 if inside area 1.0 if outside areas	Land-use raster image
	Distance to conservation area	Conservation areas, such as natural reserves and national parks	0.0 inside and thereafter linearly decaying function with distance	Land-use raster image
	Distance to residential areas	Proximity to residential areas is not desirable	0.0 inside and thereafter linearly decaying function with distance	Population raster image
Economical	Extension of engineering structures	Engineering structures such as tunnels, viaducts, and bridges, are the most expensive of all road structures	$1.0 - \frac{\text{extensions of ES}}{\text{total extension route}}$	Based upon the hydrography, rail, and road raster images
	Price of the land	Urban land is more expensive than others. We take the traversed urban land extension	$1.0 - \frac{\text{extensions of UL}}{\text{total extension route}}$	Based upon the land-use raster image

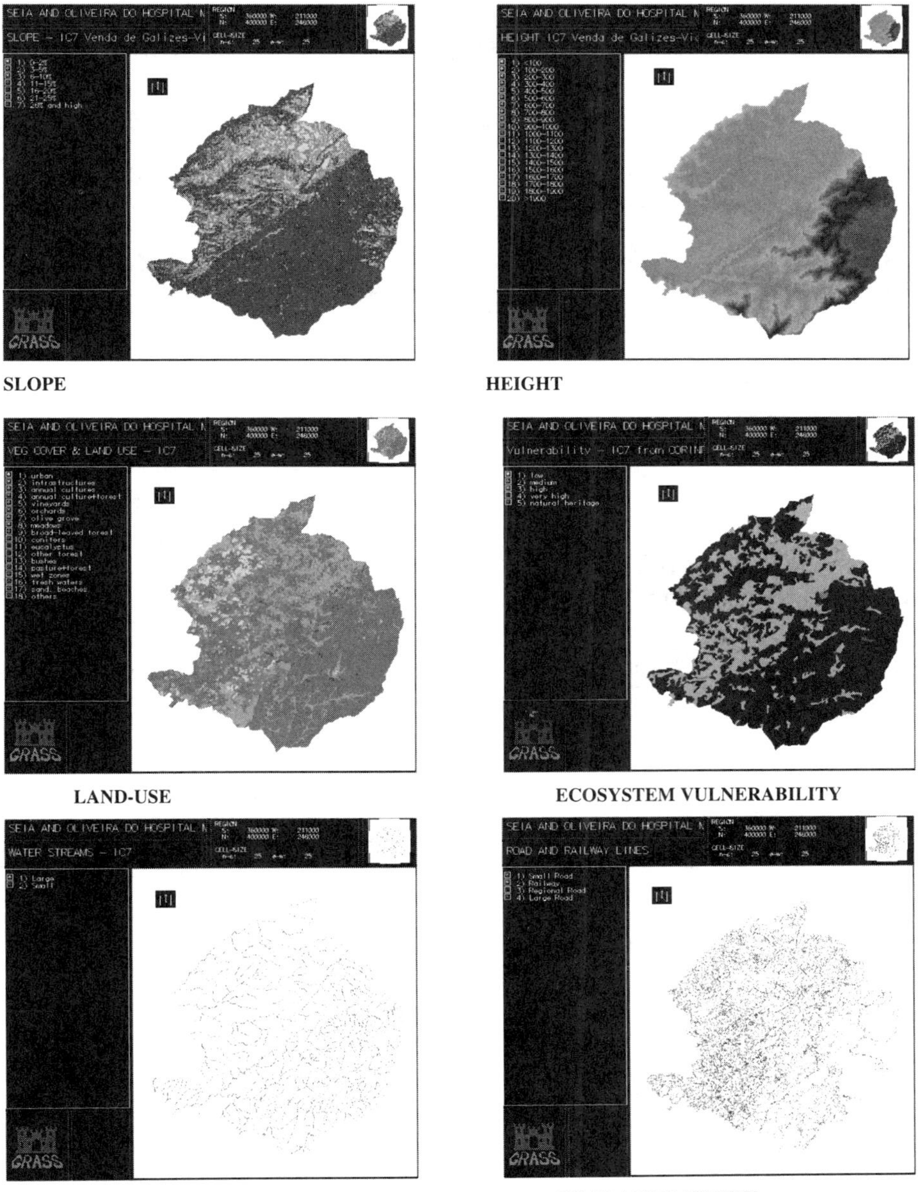

Figure 6-5 Thematic maps used in the design of routes. Source: This map was obtained from information digitized at 1 : 25,000 scale. SLOPE derived from the height map with GRASS command r.slope.aspect (CERL, 1993). Reclassified.

Motorway siting criteria have been subject to several studies and surveys (e.g., OECD, 1994; Suarez Cardona, 1989; Sankoh et al., 1993). Common criteria address issues such as the proximity of the route to residential areas, protection of environmentally sensitive areas, and area of land required.

Genetic Algorithms' Requirements

Codification of Routes

As described previously ("Modeling Route Geometry"), routes are codified as n-tuples of five parameters: a straight extension (in meters), an angle of deflection (in radians), a curvature radius (in meters), a curve extension (in meters) and the sense of curvature (0 if anticlockwise and 1 if clockwise). For a design speed of 120 km/h and a resolution[6] of 250 m, the number of bits in a binary string is 23 per tuple. The analysis of these tuples is referred to the starting point and to the end of the route. The number of tuples and therefore the number of bits in the string depends upon the distance between the target points. Typically the Euclidean distance between the two target points is calculated assuming that the total length of the route could not exceed twice the rectilinear distance between the two target points. Thus, the string will vary accordingly, giving rise to different string lengths. The Euclidean length will indicate the number of necessary tuples to link the two points.

Fitness Function

The evaluation phase of the route case and in general of facilities that are spatially represented by more than one cell or that traverse several cells had to be slightly modified to aggregate for each criterion the score values of the cells. As a strength value, Φ is used, which, as explained earlier, is based on the arithmetic mean of the criteria scores traversed by the route.

Design of GA Parameters

Population Sizing Population size is set to be small according to theory and empirical results. Some testing was performed and it was seen that good quality results were attained using populations of 50 routes for the problem size being addressed here.

Cross-over and Mutation Rates To ensure good mixing of building blocks (BBs), the one-point cross-over is set to a high value. The cross-over rate is set to 1.0 as the default value because better results were achieved on the test-bed where this was applied. Mutation is kept at a very low value, $1/N$, an empirically widely applied mutation rate. The specification of these parameters is done in table 6-5 for the case study addressed here.

Table 6-5 Genetic algorithm parameters for the IC7 route

Parameters	Type or Range	Preferred and Adopted Values
Population size	50–100	50
Selection scheme	Tournament with and without replacement	TS without replacement
Recombination type	One-point and two-point cross-over	One-point cross-over
Cross-over rate	0.75–1.0	0.85
Bitwise mutation rate	$1/N - 1/l$	$1/N$
Selection strategy	Niching, elitist	Niching, elitist

Niching Settings: σ_{share} *and* $d(i, j)$ When two alternative routes participating in a tournament do not show a clear dominance relation, fitness sharing is applied (Goldberg & Richardson, 1987; Oei et al., 1991; Mahfoud, 1994, 1995).

The distance function $d(i, j)$, between two individuals is a function of the percentage of common cells traversed by the two routes. Euclidean distances between each pair of target coordinates, from the starting coordinates to the ending coordinates, are calculated and then the percentage of common traversed cells is computed. If the number of cells traversed by the two routes is different, $d(i, j)$ is 1.0, indicating that the routes do not pertain to the same niche. σ_{share} is the maximum allowed percentage of noncoincident cells for two routes to pertain to the same niche. Values of 0.1 to 0.5 (percentage of cells) were tried out, and eventually σ_{share} is set to 0.1.

The sharing function, $Sh(d(i, j))$, is computed for the individuals in the population being filled (continuous update sharing) and that have the same fitness function value, using the distance $d(i, j)$. The niche count is calculated as

$$\Psi(i) + \sum_{j=1}^{N} Sh(d(i, j))$$

Search Results Using Evolutionary Algorithm

Simulation Results for a Set of Criteria

Characterization of the Experiments

The multicriteria evaluation function applied in these experiments is ranking based, as explained. The GA parameters l and N are set on-line as already explained; the cross-over and mutation rates were set as in table 6-5, which summarizes the GA parameters used in these experiments.

Experimental Results

Figure 6-5 is the result of applying this methodology to the set of criteria presented in table 6-4. In table 6-6 the strongest five alternative routes obtained are

Table 6-6 Criteria scores of the final alternative routes, for the criteria listed in table 6-4 in that order. These routes are displayed in figure 6-6 and numbered for easy identification

Routes	C1	C2	C3	C4	C5	C6	C7	C8	C9	C10	C11
1	0.43434	0.52319	0.45702	0.98421	0.82631	0.97979	0.61377	0.46858	0.97233	0.81027	0.44345
2	0.77778	0.62062	0.44641	0.96650	0.85327	0.54040	0.3685	0.44778	0.97536	0.82758	0.47891
3	0.55556	0.34203	0.45054	0.97849	0.83871	0.60606	0.7262	0.44415	0.96981	0.83018	0.15647
4	0.13131	0.7662	0.42617	0.99012	0.84938	0.90909	0.6399	0.42807	0.95906	0.85672	0.15574
5	0.62435	3.06249	0.44193	0.98064	0.83871	0.39393	0.32983	0.45177	0.97494	0.81837	0.55037

characterized for each of the criteria. Tables 6-7a–6-7e describe the numbered routes of figure 6-6 for their plan view geometric parameters, and length and number of engineering structures to be built if water streams are crossed. Note that by changing criteria scores or by adding other criteria to the set it is possible to generate different alternatives.

Table 6-7(a) Parameters for Route 1

R1	T1	T2	T3	T4
SE	1600	1000	2400	2000
SA	0.2	4.7	2.6	0
CE	0	300	600	0
CR	5900	1000	0	0
S	0	0	0	0
Length	2400	1450	3000	2000

Table 6-7(b) Parameters for Route 2

R1	T1	T2	T3	T4
SE	1200	0	2245	300
SA	0	5.5	0.3	0
CE	0	3400	500	0
CR	2500	3100	0	
S	0	0	0	0
Length	1200	5000	4355	300

Table 6-7(c) Parameters for Route 3

R3	T1	T2	T3	T4
SE	2200	300	2500	1660
SA	0.4	0.2	0.3	0
CE	600	0	250	400
CR	3000	0	2500	2300
S	1	0	0	0
Length	3800	300	500	2500

Table 6-7(d) Parameters for Route 4

R4	T1	T2	T3	T4
SE	2500	400	1000	1700
SA	0.4	0.2	0.3	0
CE	400	0	520	400
CR	3500	0	2600	2300
S	1	1	0	0
Length	5000	500	2500	2700

Table 6-7(e) Parameters for Route 5

R5	T1	T2	T3	T4
SE	2350	350	1250	1880
SA	0.7	0.5	0.6	0.7
CE	400	1500	250	400
CR	3700	2300	2500	2300
S	1	0	1	1
Length	4000	300	4500	2500

R*n*: route number *n*; T*n*: tuple number *n*; SE: straight extension; SA: straight angle of deflection; CE: curve extension; CR: curve radius; S: curve deflection.

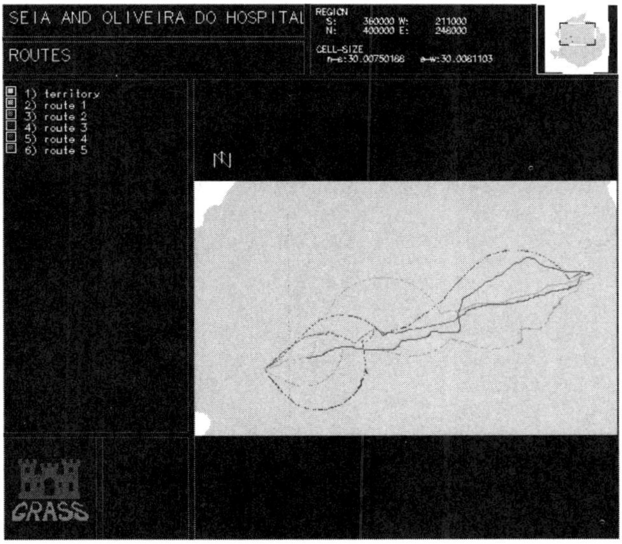

Figure 6-6 Final population of routes after 10 generations. Characteristics for the *stronger* routes are shown in tables 6-6 and 6-7. Routes are numbered for easy identification.

Conclusions

General Conclusions

In the previous section it was seen that the GA search engine delivers routes that are geometrically coherent according to the modeled geometric parameters. This route optimization simulation provided evidence that the GA is competent, as expected, as the optimization algorithm for route design, since it returned several compromise combinations of geometric parameters contained in the route specifications, with equivalent fitness.

Adding georeferenced environmental criteria in the analysis and therefore constraining the development of the route in the terrain according to points of view other than technical has been shown to lead to equally reasonable results. These results indicated that the GA-MCE-based technique is promising for comprehensive design of road routes.

Compared with other route design algorithms (briefly discussed above in "Route Geometry of Roads") the GA-MCE implemented in a GIS has the following advantages:

- it incorporates the geometry of the route—through the codification mechanism of the GA;
- it allows the analysis of the corridor traversed by the route for georeferenced information—the GIS framework facilitates this task;
- it allows multiple solutions to be compared for multiple criteria—in the GA, a population of alternatives is generated and the fitness each alternative can be compared with others by means of multicriteria evaluation techniques;

- it allows multicriteria evaluation of the various alternatives—it makes use of semicompensatory strategies;
- visualization of the routes in the terrain—the GIS framework provides the means for reporting and graphical display of the generated routes.

Finally, the main disadvantage of this technique is the GA parameter setting, which might require skills that the ordinary user of GIS does not have. To overcome this issue a large effort has been put into both empirical and theoretical setting of parameters, in particular population size. The other GA parameters are more easily set and default values are introduced in the GA. Of course, these values can be changed by the user if so wished.

Further Research

Validation of the Geometric Model

The application of the geometric algorithm needs field validation. However, it was demonstrated that GAs can be valuable to generate alternative routes, no matter what geometric requirements are included in the analysis. The GA acts independently of the problem—a *black box* algorithm—and all it requires is a measure of fitness of the individuals it is processing. Therefore, it is believed here that the validation of the geometric model in the terrain will enhance the quality of the resulting routes. Validation of the geometric model will probably require more programming or even different codification of the routes. That is an issue for further research.

Inclusion of Other Points of View

Although it is desirable to have had more criteria for other points of view to evaluate the candidate routes, the analysis performed here had to be based on the available data. The usefulness of this technique is very much dependent on the availability of georeferenced data to build criteria and on the resolution of the digitized data which enhances the quality and accuracy of the generated alternative routes if large.

Acknowledgments The author wishes to acknowledge Junta Nacional de Investigação Científica e Tecnológica, Portugal, for the financial support of this work, through the contract BD/1002/90 of the Programa CIÊNCIA and contract PRAXIS XXI BD/2632/94.

Notes

1. Efficiency in the sense of the goodness for the problem at hand.
2. N.B. the total deflection is relative to the axis that links the target coordinates.
3. Alternatives in the sense of potential actions, not implicitly feasible.
4. In the GA terminology this value is the fitness.
5. By suitability, it is meant that for the criterion considered, there is a value or a range of values that are good or acceptable for the problem at hand.
6. The number of bits depends upon the resolution of the maps used in the analysis. See discussion in this section.

References

Bowyer, A. & J. Woodwark. 1983. A Programmer's Geometry. London: Butterworths.

Brans, J.P. & B. Mareschal. 1990. The PROMETHEE methods for MCDM; The PROMCALC, GAIA and BANKADVISER software. In: C.A. Bana e Costa (ed.), Readings in Multiple Criteria Decision Aid, 216–252. Berlin: Springer-Verlag.

Brans, J.P., B. Mareschal & Ph. Vincke. 1986. How to select and how to rank projects, The PROMETHEE method. European Journal of Operational Research, 24, 228–238.

CERL, U.S.A. 1993. GRASS Version 4.1: User's Reference Manual, Open GRASS Foundation.

Fonseca, C.M. & P.J. Fleming. 1993. Genetic Algorithms for Multiobjective Optimization: Formulation, Discussion and Generalization. In: Fifth International Conference on Genetic Algorithms, Pennsylvania, Morgan Kaufmann.

Goldberg, D.E. & J. Richardson. 1987. Genetic algorithms with sharing for multimodal function optimization, Second International Conference on Genetic Algorithms, Cambridge, Mass. Mahwah, N.J.: Lawrence Erlbaum Associates, 41–49.

Guimarães Pereira, A. 1996. Generating alternative routes by multi-criteria evaluation and a genetic algorithm. Environment and Planning B: Special Issue on Connectivity Computing, 23.

Guimarães Pereira, A. 1997. Extending Environmental Impact Assessment Processes: Generation of Alternatives for Siting and Routing Infrastructural Facilities by Multicriteria Evaluation and Genetic Algorithms. New University of Lisbon. PhD Thesis.

Guimarães Pereira, A. & P. Antunes. 1996. Extending the EIA Process: Generating Alternatives for Routing. IAIA 96, Estoril.

Horn, J. & N. Nafpliotis. 1993. Multiobjective Optimization Using the Niched Pareto Genetic Algorithm. Urbana-Champaign, University of Illinois.

JAE, J. A. d. E.-M. d. O. P. 1992. Normas de Traçado—Elementos Básicos: Norma P2-91. Lisboa, JAE-MOPTC, 76.

Janssen, R. & P. Rietveld. 1990. Multicriteria analysis and geographical information systems: and application to agricultural land use in the Netherlands. In: H.J. Scholten & J.C.H. Stillwell (eds.), Geographical Information Systems for Urban and Regional Planning, 129–139. Dordrecht: Kluwer Academic.

Kanayama, Y. & N. Miyake. 1986. Trajectory generation for mobile robots. Robotics Search, 3, 333–340.

Keeney, R.L. 1992. Value-Focused Thinking. Cambridge, Mass.: Harvard University Press.

Keeney, R. & H. Raiffa. 1976. Decision with Multiple Objectives: Preferences and Value Trade-offs. New York, Wiley & Sons.

Mahfoud, S.W. 1994. Population sizing for sharing methods. IlliGAL Report no. 94005. Urbana-Champaign, University of Illinois.

Mahfoud, S.W. 1995. Niching methods for genetic algorithms. IlliGAL Report no. 95001. Urbana-Champaign, University of Illinois.

Nelson, W. 1989. Continuous-curvature paths for autonomous vehicles. IEEE International Conference on Robotics and Automation, 8, 1260–1264.

OECD 1994. Environmental Impact Assessment of Roads. Paris: OECD.

Oei, C.K., D.E. Goldberg, & S.-J. Chang. 1991. Tournament selection, niching, and the preservation of diversity. IlliGAL Report no. 91001. Urbana-Champaign, University of Illinois.

Rowe, N.C. & R.F. Richbourg. 1990. An efficient Snell's law method for optimal-path planning across multiple two-dimensional, irregular, homogeneous-cost regions. The International Journal of Robotics Research 19(6), 427–436.

Roy, B. & P. Vincke. 1981. Multicriteria analysis: survey and new directions. European Journal of Operational Research, 11(8), 207–218.

Sankoh, O.A., D. Bonner, T. Determann, J. Gersten, A. Groppe, I. Hoelters, N. Krueger, U. Lehmann, C. Marty, K. Strauss, & H. Wegner. 1993. Finding and assessing route alternatives. Journal of Environmental Management, 38, 323–334.

Suarez Cardona, F. 1989. Guías Metodológicas para la Elaboración de Estudios de Impacto Ambiental: 1. In: S. Gonzalez Alonso & J.I. Gamarra Rocandio (eds.), Carreteras y Ferrocarriles, 165. Madrid: MOPU.

Vanderpooten, D. 1990. The construction of prescriptions in outranking methods. In: C.A. Bana e Costa (ed.), Readings in Multiple Criteria Decision Aid, 184–215. Berlin: Springer-Verlag.

Vansnick, J.-C. 1990. Measurement theory and decision aid. In: C.A. Bana e Costa (ed.), Readings in Multiple Criteria Decision Aid, 81–100. Berlin: Springer-Verlag.

Wienke, D., C. Lucasius, et al. 1992. Multicriteria vector optimization of analytical procedures using a genetic algorithm. Part I: Theory, numerical simulations and application to atomic emission spectroscopy. Analytica Chimica Acta, 265, 211–225.

7

Airspace Sectoring by Evolutionary Computation

DANIEL DELAHAYE

When joining two airports, aircraft must follow routes and beacons; these beacons are necessary for pilots to know their position during navigation and because of the small number of beacons on the ground they often represent crossing points of different airways. Crossing points may generate conflicts between aircraft when their trajectories converge on it at the same time and induce a risk of collision.

At the dawn of civil aviation, pilots solved conflicts themselves because they always flew in good weather conditions (good visibility) with low-speed aircraft. In contrast, modern jet aircraft do not enable pilots to solve conflicts because of their high speed and their ability to fly with bad visibility. Therefore, pilots must be helped by an air traffic controller on the ground who has a global view of the current traffic distribution in the airspace and can give orders to the pilots to avoid collisions.

As there are many aircraft simultaneously present in the sky, a single controller is not able to manage all of them. Airspace is then partitioned into different sectors, each of them being assigned to a controller.

Sectoring is currently done in an empirical way by some airspace experts who apply rules they have learned with experience. The sectoring modifications are usually due to traffic evolution over long period and when a sector is regularly overloaded it has to be modified. To reach this aim, an ad hoc commission meets to identify new boundaries for the sectors in order to balance the workload. Afterward, sectoring is updated (until new problems arise).

This way of working is relevant because it takes into account several practical aspects but has a limited effect on the local zone it treats. This process can be improved with an automatic approach in order to give a solution to the sectoring problem in the whole airspace and that solution could be refined by experts.

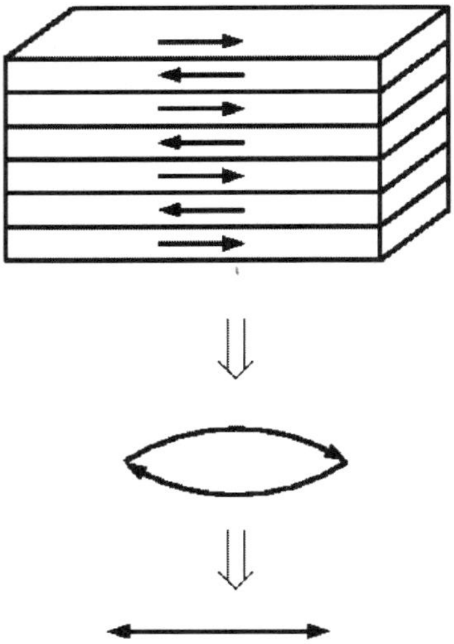

Figure 7-1 Airways modeling.

Before specifying a mathematical description of our problem, it is necessary to set out our framework to introduce some simplifications for our model. Since the training period of an air traffic controller on his or her sector is long (from 3 to 4 months), real-time sectoring optimization according to the variations of the traffic load has not been investigated. Instead, a registered maximum load traffic period on the working network has been considered. Our problem is then to partition the airspace in order to generate a balanced control workload over all sectors.

When examining the physical air traffic network, we notice that airways are superposition of several routes which have the same projection on the ground but different altitudes according to their azimuth.[1] So an airway can be modeled by a bidirectional link that gathers several individual aircraft routes (see figure 7-1). Then, our three-dimensional transportation network will be modeled by a classical two-dimensional network on a horizontal plan.

When aircraft go from point A to point B, they have to use airways of the air traffic network like drivers do on the road network. As on a road, there are crossings in the sky and aircraft have to safely pass these points. Because of their speed, aircraft cannot make anticollision procedures alone and are helped by controllers who solve the different conflicts that may arise. But nowadays, there are too many flights in the airspace (for instance an average of 6000 aircraft movements is registered everyday in the French airspace) for a single controller to manage all this traffic; thus airspace is divided into several sectors, each sector being assigned to a controller. Like any human being, a controller has working limits and a sector is said to be saturated when there are three conflicts to solve and 15 aircraft in that sector (this rate must not last too long).

The controller workload has several sources that can be divided into two categories:

- there are quantitative factors (the number of flights, number of conflicts etc.), which can be precisely modeled in a mathematical way and handled by an optimization algorithm;

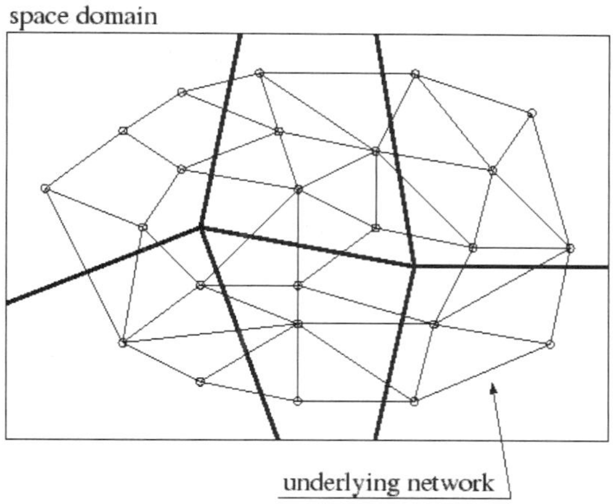

Figure 7-2 Example of sectoring.

- there are psychological factors (stress, concentration etc.), which have no evident mathematical formulation but are in direct relationship with the previous ones according to the controllers themselves.

So, only quantitative elements will be taken into account in our application on first approximation. Having a model, our goals can be define more precisely in the following way:

> Consider an air traffic transportation network in a two-dimensional space with flows on it inducing a workload distributed over the space. This workload must be partitioned into K balanced sectors in order to minimize coordinations (when an aircraft crosses a sector boundary, controllers in charge of those sectors have to exchange information about the flight, inducing a workload called coordination).

Figure 7-2 shows an example of network sectoring with six sectors. This sectoring must take some constraints into account coming from the air traffic control system:

- A pilot must not encounter the same controller twice during his or her flight to prevent superfluous coordinations; this means that an aircraft crossing a sector will encounter two and only two sector boundaries. To guarantee that our sectors meet this constraint we force them to be convex in the topological sense.[2] This constraint gives sectors a polygonal shape.
- A sector boundary has to be at least at a given distance from each network node (security constraint). When a controller has to solve a conflict, he or she needs a minimum amount of time to develop a solution. Each controller manages his or her sector individually, if a sector boundary is too close to a crossing point, the controller is not able to solve any conflicts because there is not enough time between the coordination step (with the previous sector where the aircraft comes from) and the time the aircraft reaches the crossing point. The minimum delay time is fixed at 7 minutes and can be converted into a distance once the aircraft speed is known.

- An aircraft has to stay at least a given amount of time (a few minutes) in each sector it crosses to give enough time to the controller to manage the flight in optimum conditions (minimum stay time constraint). This constraint is represented by a minimum distance between two boundaries cutting the same network link. The two last constraints will be implemented the same way by forcing a minimum length for any link segment between two consecutive boundaries or between a node and a boundary.

In the next section, a more precise description of our problem is given and some simplifications are done in order to develop a mathematical model. The third section presents a continuous approach (Delahaye et al., 1994) based on Voronoi diagrams to solve this partitioning problem. The last section presents a discrete approach (Delahaye et al., 1998).

Problem Modeling

A transportation network is defined as a doublet (N, L) in which N is the set of nodes (with their positions in a topological space) and L is the set of links, each of them transporting a quantity f_{ij} of flow from node i to node j (Klingman & Mulvey, 1981). This original network supports a control workload related to the link flows, and is built by applying the following steps:

- A loaded day of traffic is chosen.
- All the crossings between all the trajectories are registered.
- The associated workload is computed for this network.
- An initial network is created where all the nodes represent a crossing (this network is supposed to be connected, this means that each node of the network can be reached from all the other nodes by an undirected path).
- All the links greater than 14 minutes are removed and saved in a new set of links.
- The remaining connected components (and their associated workloads) are gathered into a new single node (a connected component is a connected subgraph).
- A new contracted network is created from the two new sets of nodes and links.

This new network will be partitioned into sectors. As stated before, we just take quantitative criteria into account to compute controller workload (Tuan et al., 1976). According to the controllers themselves, workload can be divided into three parts, which represent respectively the conflict workload, the coordination workload, and the trajectory monitoring workload of the different aircraft that are present in a sector:

- The conflict workload (C_{cf}) gathers the different actions of the controller to solve conflicts.
- The coordination workload (C_{co}) represents the information exchanges between a controller and the controller in charge of the bordering sector or between a controller and the pilots when an aircraft crosses a sector boundary.
- The monitoring workload (C_{mo}) aims at checking the different trajectories of the aircraft present in a sector.

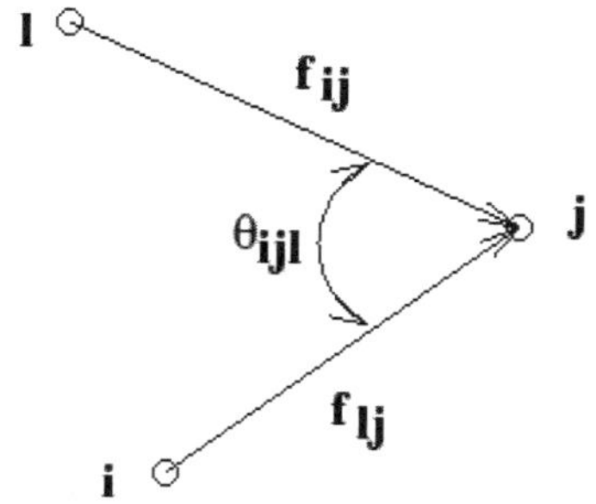

Figure 7-3 Route crossing.

Conflict Resolution Workload

Two aircraft are said to be in conflict when their relative distance reaches a lower bound called separation standard. It can be shown that the conflict workload on a crossing of two airways can be estimated by the formula (figure 7-3):

$$C_{cf} = \alpha(\theta_{ijl}) f_{ij} f_{lj}$$

where θ_{ijl} is the angle between links (i, j) and (l, j), $\alpha(\theta_{ijl})$ is a weighted coefficient.

For a crossing with more than two airways:

$$C_{cf}(j) = \sum_{\substack{i \in N \\ i \neq j}} \sum_{\substack{l \in N \\ l \neq i \\ l \neq j}} \alpha(\theta_{ijl}) f_{ij} f_{lj}$$

And for a sector S_k, the associated conflict workload is the summation of the conflict workloads over all the crossing points belonging to the sector S_k:

$$C_{cf}(S_k) = \sum_{j \in N_k} \sum_{\substack{i \in N \\ i \neq j}} \sum_{\substack{l \in N \\ l \neq i \\ l \neq j}} \alpha(\theta_{ijl}) f_{ij} f_{lj}$$

N_k is the number of nodes in the sector S_k.

For instance, the conflict workload of the sector S_k represented in figure 7-4 is given by:

$$\begin{aligned} C_{cf}(S_k) = 0.5[&\alpha_{61} \cdot (\theta_{615}) \cdot 0.5 \cdot f_{61} \cdot f_{51} \\ &+ \alpha_{12} \cdot (\theta_{127}) \cdot 0.5 \cdot f_{12} \cdot f_{72}] \end{aligned}$$

Coordination Workload

For a sectorized air network, the coordination workload is related to the flows cut by the sector boundaries. If a link (i, j) belonging to sector S_k is considered [this means that a part of the link (i, j) is in the sector S_k], three cases can be identified:

- Both extremities are in sector $S_k \Rightarrow i \in N_k$ and $j \in N_k$; all the links belong to the sector S_k [link (2, 3) in figure 7-5].
- Both extremities are in sector $S_k \Rightarrow i \in N_k$ and $j \in N_k$; all the links belong to the sector S_k [(link (2, 3) in figure 7-5].

$$\Rightarrow C_{co} = 0$$

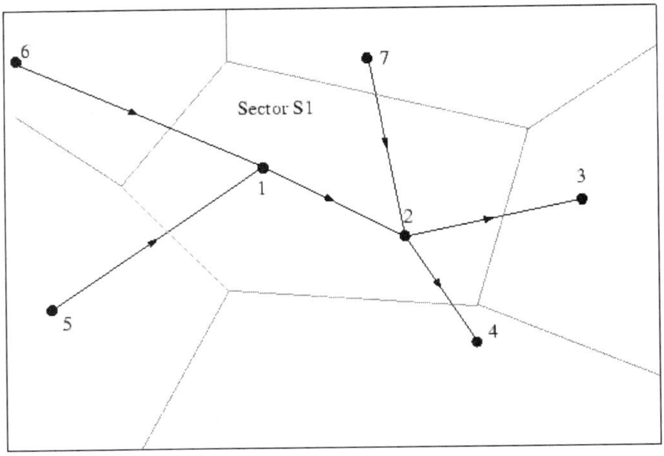

Figure 7-4 Typical control sector.

- Only one extremity is in the sector $S_k \Rightarrow i \in N_k$ or (exclusive) $j \in N_k$. There is an intersection between the link and the sector boundary [links (1, 2), (3, 4), (6, 2), (3, 5) in figure 7-5]. In this case:

$$\Rightarrow C_{co} = \beta_{ij} f_{ij}$$

where β_{ij} is a weight coefficient.

- Both extremities are outside the sector but the intersection between the sector and the link is not empty:

$$\Rightarrow i \notin N_k \quad \text{and} \quad j \notin N_k \quad \text{with} \quad (i, j) \in L_k$$

where L_k is the subset of a link having a nonempty intersection with the the sector S_k. This means that link flow f_{ij} is cut twice [link (1, 4) in figure 7-5]:

$$\Rightarrow C_{co} = 2\beta_{ij} f_{ij}$$

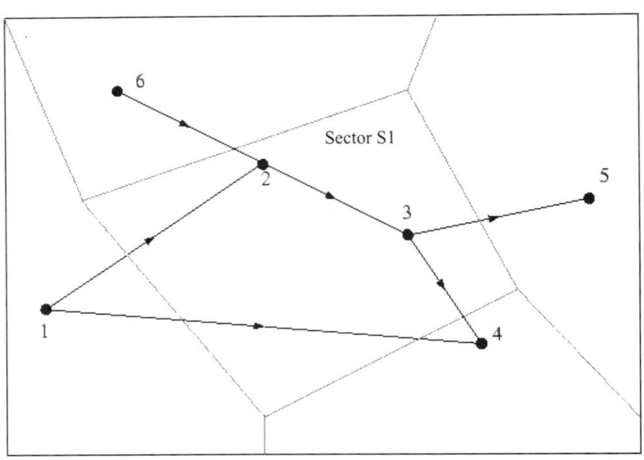

Figure 7-5 Network with three kinds of coordination.

The global coordination workload associated with the sector S_k is given by

$$C_{co}(j) = \sum_{\substack{i \in N_k \\ \oplus j \in N_k}} \beta_{ij} f_{ij} + \sum_{\substack{i \notin N_k \\ j \notin N_k \\ (i,j) \in L_k}} 2\beta_{ij} f_{ij}$$

where $\oplus$ means "exclusive or."

In the example: $C_{co}(S_1) = \beta_{12} f_{12} + \beta_{62} f_{62} + \beta_{35} f_{35} + \beta_{34} f_{34} + 2\beta_{14} f_{14}$.

Monitoring Workload

This workload is directly related to the number of aircraft in the sector. For a sector S_k it is given by

$$C_{mo}(S_k) = \eta \sum_{(i,j) \in L_k} p_{ij}(k) f_{ij}$$

where $p_{ij}(k)$ is the proportion of link (i, j) belonging to the sector S_k, η is a weighted coefficient.

Global Workload Induced in a Sector

The global control workload is just the summation of the conflict workload, the coordination workload, and the monitoring workload.

$$C(S_k) = C_{cf}(S_k) + C_{co}(S_k) + C_{mo}(S_k)$$

Having a model, we can now define our goals more precisely in the following way: one considers an air traffic transportation network in a two-dimensional space with flows on it inducing a workload distributed over the space. This workload must be partitioned into K balanced sectors in order to minimize coordinations. The induce sectoring has to meet the safety constraint, the route convexity constraint and the minimum stay time constraint.

If C_{gl} represents the global workload all over the airspace, the optimum balancing for a sectoring with K sectors is given by

$$C(S_k)_{opt} = \frac{C_{gl}}{K}$$

From this optimum, it is possible to build a relative balancing criterion C_E:

$$C_E = \sum_{k=1}^{K} \frac{\left| C(S_k) - \dfrac{C_{gl}}{K} \right|}{\dfrac{C_{gl}}{K}} \quad 0 \leq C_E \leq 2K - 1$$

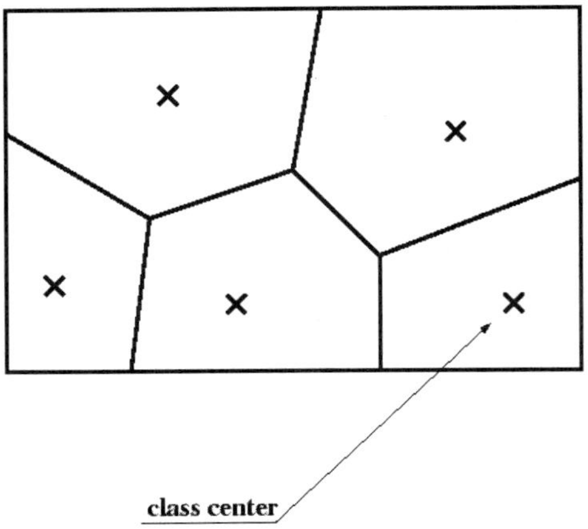

Figure 7-6 Example of a five-partitioning.

The second part of our objective function (minimization of the coordination) is then determinated by reduction of the relative proportion of the coordination workload on a sector:

$$C_C = \frac{\sum_{k=1}^{K} C_{co}(S_k)}{C_{gl(cf)} + C_{gl(mo)}}$$

Our objective is then the minimization of a weighted sum of the two previous criteria:

$$obj = \min((1 - \alpha)C_E + \alpha C_C)$$

Continuous Approach

This approach gives sectoring of the geographic airspace by assigning each point of the continuous domain to a sector. The principle of the decomposition is based on the Voronoi diagrams (Preparata & Shamos, 1985), which partition a spatial domain into convex sectors.

Construction of Sectors

According to the previous section, the sectors we have to build have to be convex (with a polygonal shape induced by the convexity property). To reach this goal, K points (the class centers) are thrown into the space domain containing the transportation network. Then all the domain points are aggregated to their nearest class center ending up with a K partitioning of our domain into convex sectors with linear boundaries. Figure 7-6 gives an example of a five-partitioning of a rectangle.

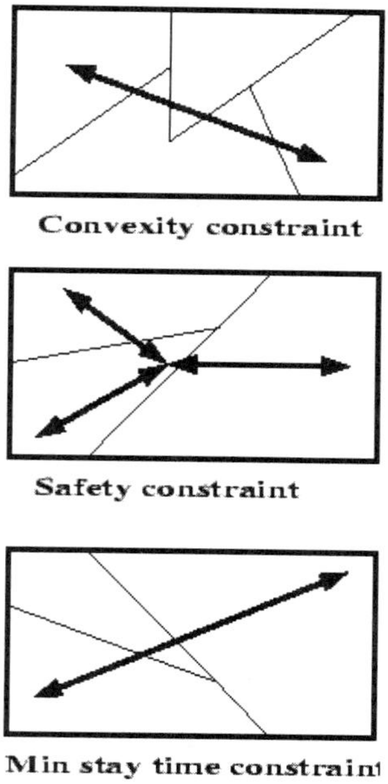

Figure 7-7 Constraint examples.

Constraints

The different constraints previously introduced are handled in the following way:

- Sector convexity: this constraint is already satisfied by the construction method of sectors.
- Security and minimum stay time constraints : these two constraints can be relaxed by an artificial increase of the coordination workload on links when the constraint is not respected:

$$C_{co}(S_k) = \sum_{\substack{i \in N_k \\ \oplus j \in N_k}} \delta \beta_{ij} f_{ij} + \sum_{\substack{i \notin N_k \\ j \notin N_k \\ (i,j) \in L_k}} 2\delta \beta_{ij} f_{ij}$$

where $\oplus$ represents exclusive or, and the penalty coefficient δ represents the constraint violation.

If the relative proportion (p_{ij}^k) of link (i, j) in the sector k is smaller than the safety distance (7 minutes of flight) then the coordination is artificially increased:

$$\text{if } p_{ij}^k L_{ij} < d_{safe} \text{ then } \delta = e^{-2\left(\frac{p_{ij}^k L_{ij}}{d_{safe}-1}\right)}$$

$$\text{else } \delta = 1$$

where L_{ij} is the length of the link and d_{safe} is the minimum distance between a crossing point and a sector boundary. Both constraints are treated the same way.

In figure 7-7 some examples are given for which the three previous constraints are not satisfied.

Complexity of Our Problem

The problem we have to solve can be divided into two separate parts corresponding to our two different goals:

1. equilibrium of the different sectors workload according to the number of aircraft and conflicts in each sector;
2. minimization of the coordination workload.

The second criterion is typically a discrete graph partitioning problem with topological constraints that is NP-hard (Garey & Johnson, 1979). Having chosen a continuous flow representation, the first criterion induces a discrete-continuous problem that is also NP-hard. So, according to the size of our network (about 1000 nodes), classical combinatorial optimization is not relevant and stochastic optimization seems to be more suitable.

Moreover, this kind of problem may have several optimal solutions (or near optimal) due to the different possible symmetries in the topological space, and so forth. The maximum number of those solutions has to be found in order to be refined by experts because we do not know at this step which one is really the best from the operational point of view. This last point makes us reject classical simulated reannealing optimization, which updates only one state variable, even if it might give better results in some cases (Ingber & Rosen, 1992).

In contrast, evolutionary algorithms (EAs) maintain and improve a population of numerous state variables according to their fitness and will be able to find several optimal (or near optimal) solutions. Then, EAs seem to be relevant to the solution of our sectoring problem.

Application of Evolutionary Algorithm to Our Problem

Evolutionary algorithms are heuristic computer search techniques whose mechanics are based upon the principles of natural selection found in the biological world (Goldberg, 1989; Back, 1996; Fogel et al., 1966; Fogel, 1994; Michalewicz, 1992; Schwefel, 1995). They have been used to obtain solutions to a diverse set of known NP-hard problems, including task scheduling, graph-theoretical problems, VLSI layout, automatic control, numerical integration, traffic assignment (Delahaye et al., 1994a; Delahaye & Odoni, 1997), and airspace sector regrouping problems (Delahaye et al., 1995). Empirical evidence strongly suggests that EAs can outperform other optimization techniques, such as simulated annealing (Aarts & Korst, 1989), mainly for multimodal objective function.

The coding consists of converting each point of the state domain into a chromosome used by the EA. It is assumed that a potential solution may be represented as a set of parameters. These parameters (known as genes) are joined together to form a string of values (often referred to as a chromosome). In genetic terms, the set of parameters represented by a particular chromosome is referred to as the genotype. The genotype contains the information required to construct an organism, which is referred to as the phenotype.

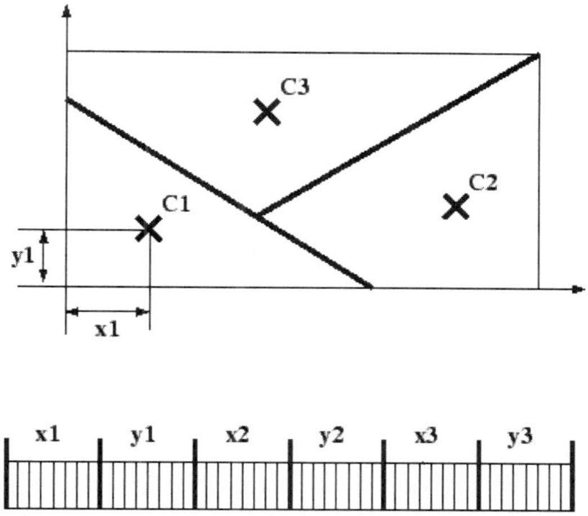

Figure 7-8 Construction of binary chromosome.

First EA: EA with Binary Chromosome

A chromosome must contain all the sectoring information for the GA to be able to evaluate the fitness for each individual. This information is summarized by a set of points in our geographical space called class centers (it can be shown that for each class center set, there is only one sectoring induced). Having chosen binary strings in this first example, the chromosome is implemented as a string of bits containing the concatenation of the different class center positions (see figure 7-8). This implementation enables us to use classical operators for GAs as shown in figure 7-9. One iteration of this binary GA is given in figure 7-10.

Fitness Evaluation

For each sector synthetized by the chromosome, the three control workloads are evaluated. Conflict workload is computed off-line on each node of the network and is assigned to sectors by a classification process during the chromosome

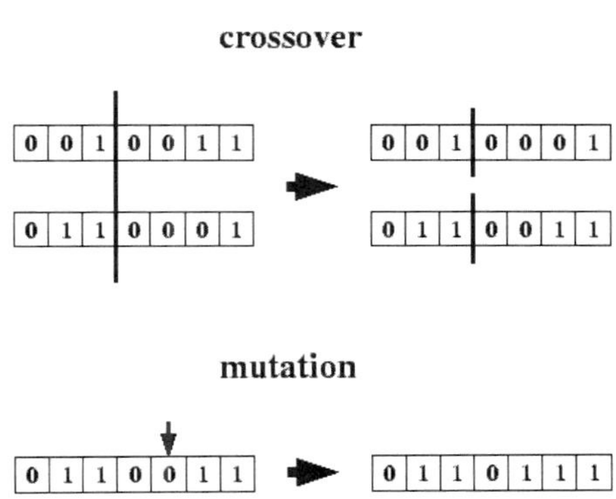

Figure 7-9 Classical binary operators.

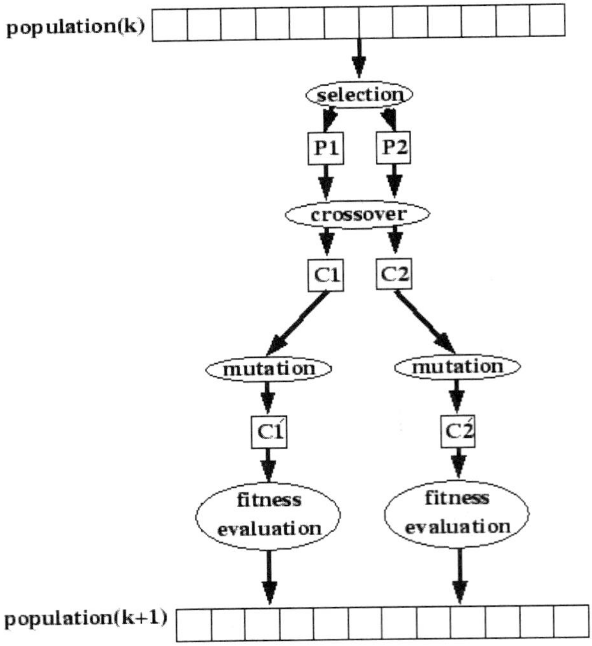

Figure 7-10 Binary GAs.

evaluation. For the two other workloads (coordination and monitoring) the following criteria have to be determined for each link of the network:

- the distribution of links over sectors;
- the number of sector boundaries cut by this link.

These two quantities are determined by dichotomy. Geographic positions of the current link nodes (segment $[N_o, N_d]$) are considered and compared with the position of the class centers (see figure 7-11). Each point of the segment $[N_o, N_d]$ is checked in order to find its closest class center.

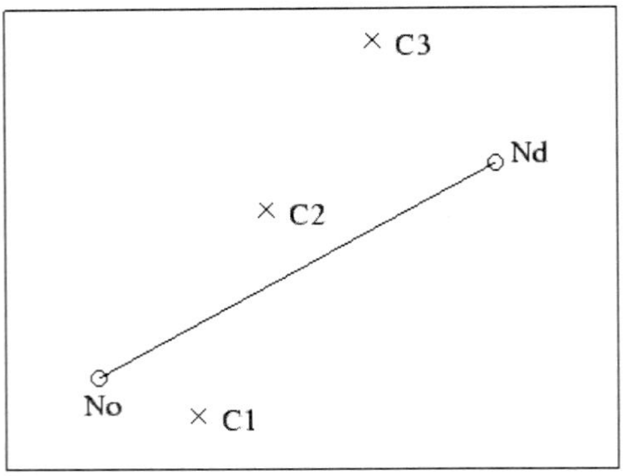

Figure 7-11 Sectoring distribution of a link.

X1	Y1	X2	Y2	X3	Y3

Figure 7-12 Floating-point chromosome.

Dichotomy Algorithm

1. $\overrightarrow{P_k}$: current point on the segment $[N_o, N_d]$, $\overrightarrow{N_o}$, $\overrightarrow{N_d}$ position vectors of the extremity nodes N_o and N_d.
2. Initialization

$$\overrightarrow{P_o} = \overrightarrow{N_o}, \overrightarrow{P_d} = \overrightarrow{N_d}, k = 0$$

3. The middle point of the segment $[\overrightarrow{P_k}, \overrightarrow{P_{k+1}}]$ is then computed:

$$\overrightarrow{P_{k+\frac{1}{2}}} = \frac{\overrightarrow{P_k} + \overrightarrow{P_{k+1}}}{2}$$

4. The closest class centers of the points $\overrightarrow{P_k}$, $\overrightarrow{P_{k+\frac{1}{2}}}$, $\overrightarrow{P_{k+1}}$ are then determined. Those points belong to their respective sectors: $S(\overrightarrow{P_k})$, $S(\overrightarrow{P_{k+\frac{1}{2}}})$, $S(\overrightarrow{P_{k+1}})$. If $\|\overrightarrow{P_{k+1}} - \overrightarrow{P_{k+\frac{1}{2}}}\| < \varepsilon$ (where ε is a parameter of accuracy) $\Rightarrow$ the boundary of sectors S_k and S_{k+1} cut the link on $\overrightarrow{P_{k+\frac{1}{2}}}$ $\Rightarrow$go to step 6; else

$$\text{if } S(\overrightarrow{P_{k+\frac{1}{2}}}) = S(\overrightarrow{P_k}) \Rightarrow \quad \overrightarrow{P_k} = \overrightarrow{P_{k+\frac{1}{2}}}$$

$$\text{else} \qquad \overrightarrow{P_{k+1}} = \overrightarrow{P_{k+\frac{1}{2}}}$$

then go to step 3;

if $S(\overrightarrow{P_{k+1}}) \neq S(\overrightarrow{N_d}) \Rightarrow \overrightarrow{P_k} = \overrightarrow{P_{k+1}}$ and $\overrightarrow{P_{k+1}} = \overrightarrow{N_d}$ then go to step 3;
else end.

Second EA: EA with Floating-Point Chromosome

In this case, each position (normalized into [0, 1]) is directly coded in a chromosome without binary conversion (Michalewicz, 1992). So, the chromosome has the structure shown in figure 7-12.

This new structure involves some new kinds of operators that we now describe.

Cross-over After selecting two parents in the current population, an allele position is randomly chosen (so two sectors are selected at the same allele position, one in each sector). Afterward, the associated class centers are joined by a straight line. Then, the class centers are moved on this line according to a uniform random variable. An example of this kind of cross-over is given in figure 7-13 (allele 1 has been selected in this example).

Mutation When a chromosome is mutated an allele position is randomly selected and its associated class center is moved by adding noise to it (it seems that best results are given with an affine distribution and not with a Gaussian one) (see figure 7-14; in this example allele 2 has been selected for mutation). The structure of this new GA is exactly the same as the binary GA in the succession of the different steps.

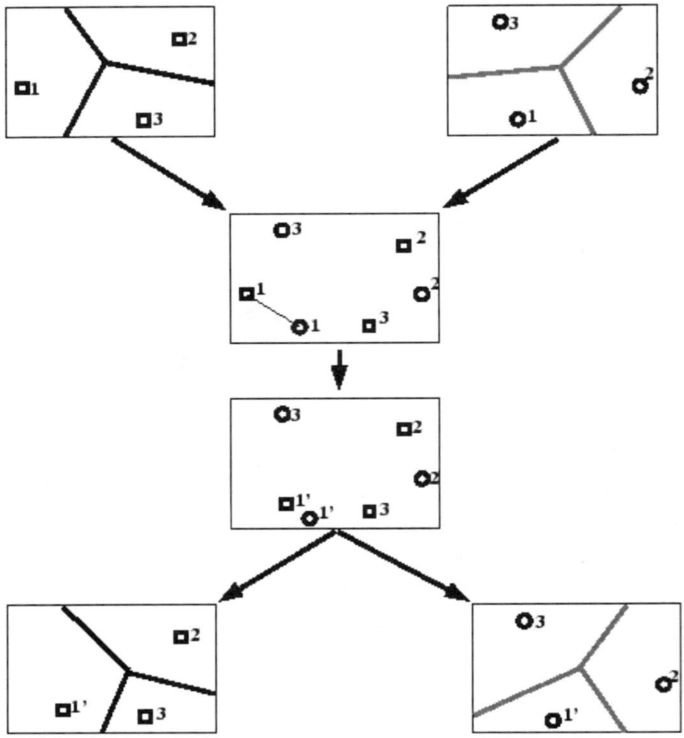

Figure 7-13 Example of cross-over.

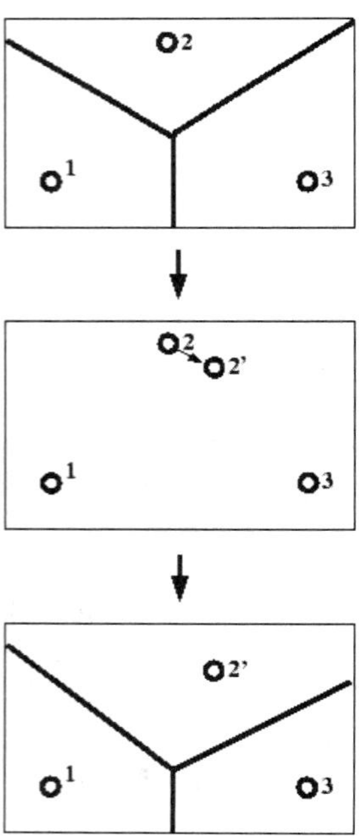

Figure 7-14 Example of mutation.

Third EA: Evolution GA

This last version of the GA is a dynamic version of the previous one with some analogy with simulated annealing. As before, parents are selected from the current population and are made to cross and mutate to create a new population.

Cross-over After selecting one class center in each parent chromosome (at the same allele position) we move each one in a random way with progressive decreasing range as the generation number increases.

This kind of cross-over process induces large movements at the beginning (⇒ quasi random exploration of the state domain) and very small ones at the end (⇒ these small movements enable the algorithm to "climb hills").

Mutation This operator randomly moves one class center in a chromosome with the same law as for cross-over. According to the extension of the distribution the initial movings are very important so space is explored in a quasi-random way and the movements become smaller as the generation number increases. The objective evaluations are refined as in a "hill climbing" process in which the climbing direction is given by the selection process.

After applying those two operators, we have four individuals (two parents P1, P2 and two children C1, C2) with their respective fitness. Afterwards, those four individuals compete in a tournament. The two winners are then inserted in the next generation. The selection process of those winners is the following:

- If C1 is better than P1, then C1 is selected, or else C1 will be selected according to a probability that decreases with the generation number.
- Then, at the beginning, C1 has a probability 0.5 of being selected, although it is worse than P1, and this probability decreases to 0.01 at the end of the process. This probability decreases in an exponential way:

$$P = e^{-\left(\frac{|f(C_i)-f(P_i)|}{c_n}\right)}$$

where c_n decreases with the generation number n.

A description of this algorithm is given in figure 7-15.

Deterministic Balancing Heuristic The previous algorithms can be improved by using the fact that "heavy" sectors have to leave weight to the "light" ones in order to balance sectors. As a matter of fact, knowing the weight of each sector, it is easy to find the heavy ones and to induce moving tendencies for the class centers in order to reduce the difference of weight between sectors (this could be done only on the conflict and the monitoring workload). For instance, if a rectangular homogeneous surface density is considered (figure 7-16), it is obvious that class centers have to be moved toward the right to balance the sectors.

To understand this deterministic balancing heuristic, a simple sectoring with two sectors is considered as in figure 7-17 (this principle has been extended to more than two sectors).

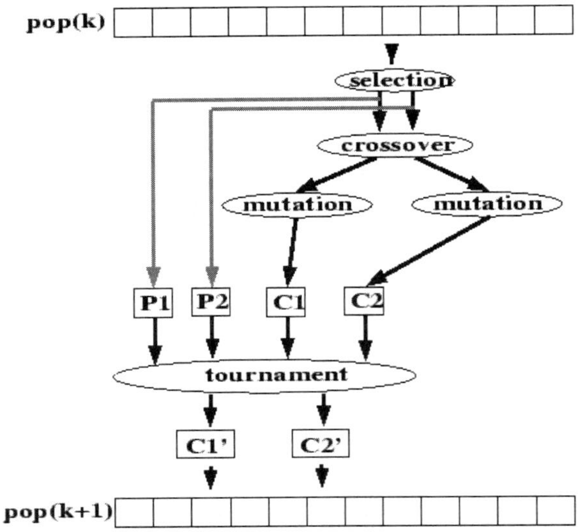

Figure 7-15 Evolution GA.

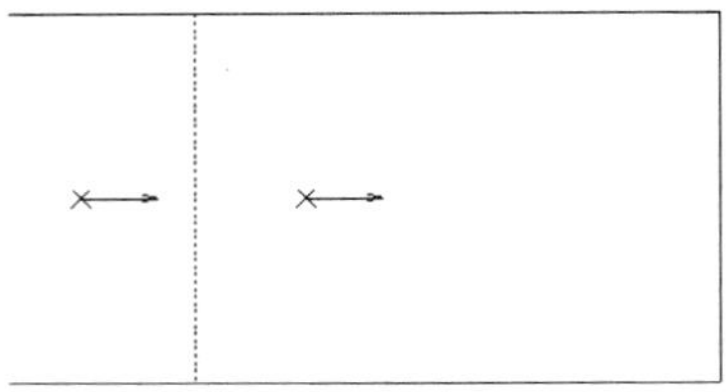

Figure 7-16 Balancing trends.

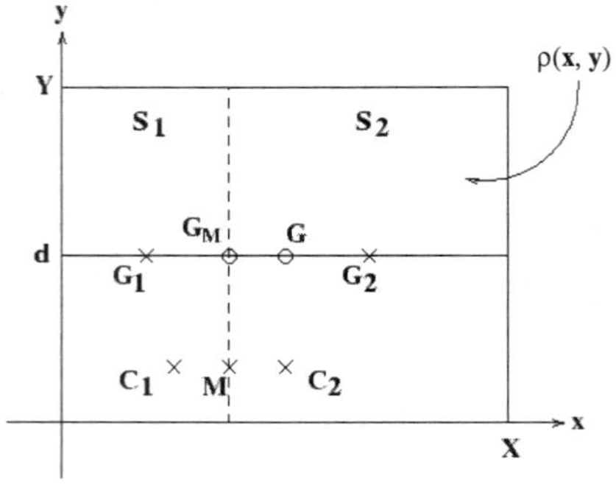

Figure 7-17 Gravity center and geometric center.

This rectangle is partitioned into two sectors by a two-class center. Without loss of generality, these class centers are supposed to be on the median line of the rectangle to simplify the expressions. The weight of these two sectors is given by:

$$m_1 = \int_0^Y \int_0^{X_M} \rho(x,y)\,dx\,dy \quad m_2 = \int_0^Y \int_{X_M}^{X} \rho(x,y)\,dx\,dy$$

where M is the middle point of the segment $[C_1, C_2]$ and $\rho(x,y)$ is the surface mass density; X and Y are the rectangular dimensions:

$$X_M = \frac{x_1 + x_2}{2}, \quad Y_M = \frac{Y}{2}$$

From these masses, gravity centers of the two sectors are computed ($\overrightarrow{G_1}, \overrightarrow{G_2}$) and the global gravity center $\overrightarrow{G}$.

The geometric center is defined by:

$$\overrightarrow{G_M} = \frac{\overrightarrow{G_1} + \overrightarrow{G_2}}{2}$$

The principle of the heuristic is then based on the following remark:

> The sectors are balanced if the geometric center coincides with the gravity center.

Then, class center moving will be done in order to reduce the distance between the geometric center and the gravity center. Two approaches have been investigated.

1. Analytical approach. If $\rho(x,y)$ is supposed to be constant on the surface:

$$\begin{cases} m_1 = & YG_{Mx} \\ m_2 = & Y(X - G_{Mx}) \end{cases}$$

and

$$G_x = \frac{m_1 G_{1x} + m_2 G_{2x}}{m_1 + m_2} = \frac{G_{Mx} G_{1x} + (X - G_{Mx}) G2x}{X}$$

We will find the moving Δ in order to balance the sectoring in one step: After moving the new positions of the class centers are given by:

$$\begin{cases} x_1' = x_1 + \Delta \\ x_2' = x_2 + \Delta \end{cases}$$

The new gravity center is:

$$G_x' = \frac{G_{Mx}' G_{1x}' + (X - G_{Mx}') G_{2x}'}{X}$$

The difference $\overrightarrow{G}' - \overrightarrow{G}_M' = 0$ for:

$$\Delta = \frac{X}{2} - G_{Mx}$$

2. Iterative approach. We define the series: $U^n = |m_1^n - m_2^n|$ where n is the current step number:

$$\vec{\Delta}^n = \vec{G}^n - G_M^n = \frac{m_1^n - m_2^n}{2(m_1^n + m_2^n)} [\vec{G}_1^n - \vec{G}_2^n]$$

We have:

$$\begin{cases} x_1^{n+1} = x_1^n + \dfrac{\Delta_x^n}{\alpha} \\ x_2^{n+1} = x_2^n + \dfrac{\Delta_x^n}{\alpha} \end{cases}$$

where α is a weighted coefficient and

$$\Delta_x^n = \frac{X - (x_1^n + x_2^n)}{4}$$

Then

$$m_1^n = Y\left(\frac{x_1^n + x_2^n}{2}\right), \quad m_2^n = Y\left(X - \left(\frac{x_1^n + x_2^n}{2}\right)\right)$$

Moreover

$$\Delta_x^n = \frac{X - (x_1^n + x_2^n)}{4} = \left(\frac{X - \dfrac{2m_1^n}{Y}}{4}\right) = \left(\frac{X - 2\left(X - \dfrac{m_2^n}{Y}\right)}{4}\right)$$

$$m_1^{n+1} = m_1^n + \frac{Y}{\alpha}\left(\frac{X - \dfrac{2m_1^n}{Y}}{4}\right) \quad m_2^{n+1} = m_2^n + \frac{Y}{\alpha}\left(\frac{X - \dfrac{2m_1^n}{Y}}{4}\right)$$

$$m_1^{n+1} = m_1^n + \frac{1}{\alpha}\left(\frac{XY - 2m_1^n}{4}\right) \quad m_2^{n+1} = m_2^n + \frac{1}{\alpha}\left(\frac{XY - 2m_2^n}{4}\right)$$

Convergence of U^n

$$U^{n+1} = U^n - \frac{1}{2\alpha} U^n$$

$$\Rightarrow U^{n+1} = \left(1 - \frac{1}{2\alpha}\right) U^n = \beta U^n$$

$$\Rightarrow U^n = \beta^n U^0$$

This represents a geometric series that is convergent for $\beta < 1$ ($\Rightarrow \frac{1}{4} < \alpha < \infty$).

When $\rho(x, y)$ is not constant over the surface, moving is induced by

$$\overrightarrow{\Delta_n} = \alpha_n(\overrightarrow{G_n} - \overrightarrow{G_M})$$

for which α_n is adapted to the current problem in a dynamical way (usually $0 < \alpha_n < \frac{1}{2}$).

For more than two sectors, this procedure is applied to each pair of sectors and the resulting moves are given by a weighted summation of all the individual moves.

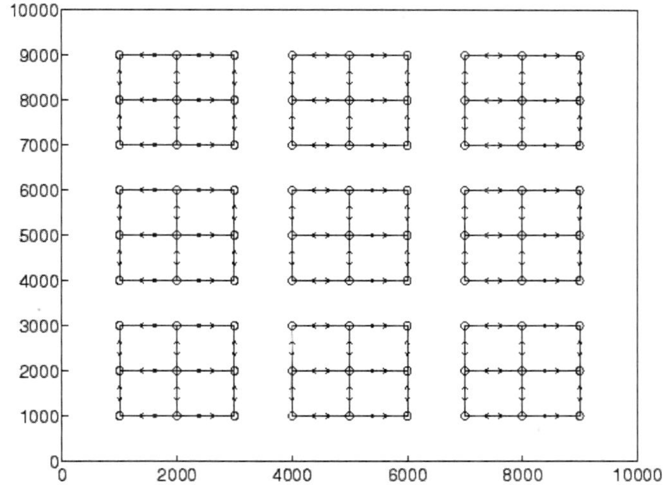

Figure 7-18 Square network.

For our problem, this heuristic has been randomly applied with a small number of iterations in order to avoid premature convergence on the first part of the criterion (workload balancing).

Results

Binary EA

These evaluations were done with the classical SGA of Goldberg (1989; Smith et al., 1991) and give good results on very small networks.

When the network size increases, this algorithm becomes inefficient because of the cross-over and mutation operators, which induce a quasi-random movement in the state space. This is due to the fact that these operators do not take into account the space point position in the chromosome and break it very roughly. This last point made us change the structure of our chromosome into a floating string where the crossing position respects each individual floating allele.

Floating EA (FGA) and Evolution EA (EGA)

The previous algorithm being too limited, we tried and compared two floating-point EAs (FGA, EGA). The results of those two algorithms are very encouraging as shown by the following experiments. To compare and evaluate these algorithms, artificial test networks have been used (see figures 7-18 and 7-19).

As we can see the first network has trivial solutions with nine sectors. These solutions seem to be very evident to a human because of the brain's ability to perceive the different symmetries, but for a computer these problems have no characteristic and remain difficult.

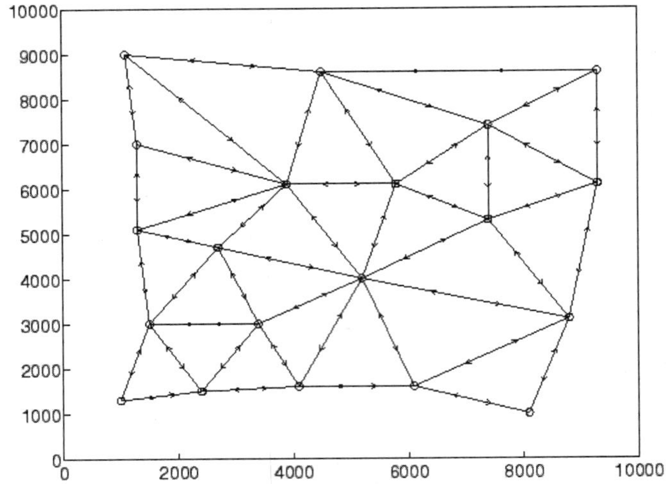

Figure 7-19 Asymmetric network.

The different parameters used for our experiments are the following:

- population size: 400
- number of generations: 200
- probability of crossover: 0.6
- probability of mutation: 0.06

Convergence

To see the convergence of our algorithms we observed the evolution of the population statistics (maximum and average) over the generations. As we can see in figure 7-18, the "square network" can be partitioned into nine sectors in several ways and will be easier to manage as it has not a unique solution but a solution set. Both algorithms find an exact solution very quickly (15 generations for FGA and eight for EGA, see figures 7-20 and 7-21), but EGA gives better results on the average statistics.

After this first experiment, the algorithm was used to partition an asymmetric network into five sectors (see figure 7-22). This network has no trivial solutions (nor does our future real air network) and it seems evident that the solution space is much smaller than for the "square network." This last point induces a slower convergence rate (as we can see in figures 7-22 and 7-23). The fitness does not reach 1 because it is not possible to partition this network without cutting link flows.

According to the balance error results, the given solutions are very close to our objective and the most unbalanced sectors are less than 0.7 percent distant from the objective. The physical sectoring result satisfies the topological constraints, as we can see in figure 7-24.

The strong point of this continuous approach is its ability to partition airspace even if aircraft move out of the regular network of airways. If aircraft stay on this kind of network, a discrete approach can be used which gives the same quality of results in a shorter time.

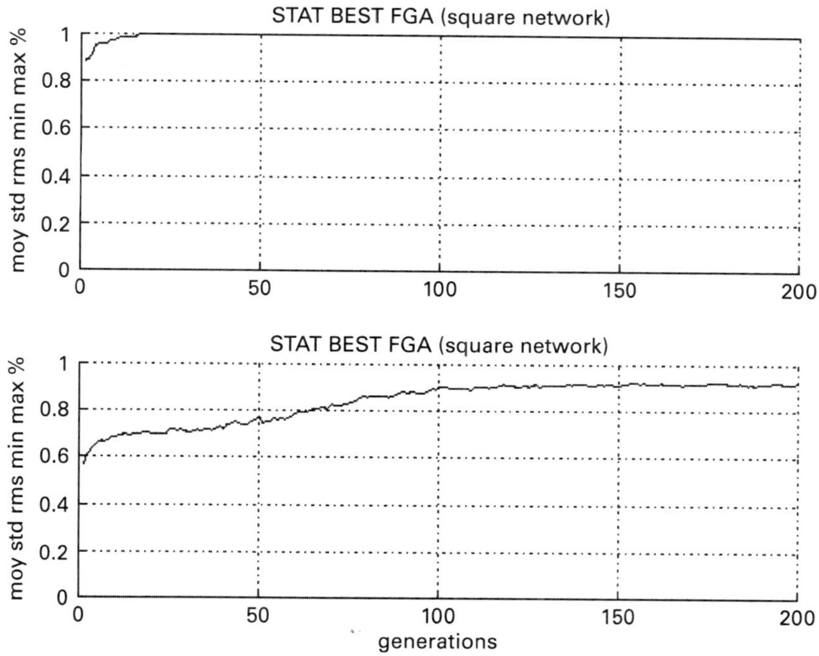

Figure 7-20 FGA stat results for the square network.

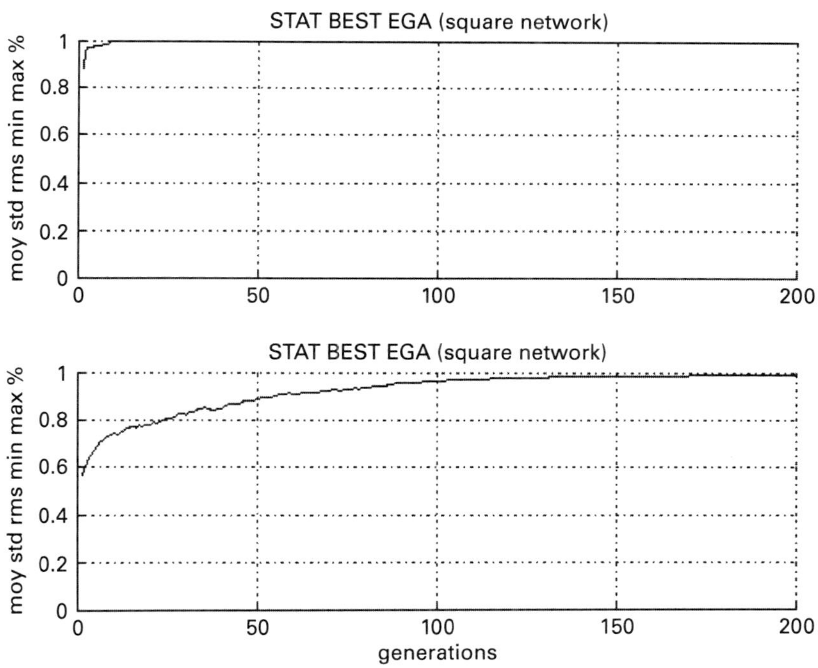

Figure 7-21 EGA stat results for the square network.

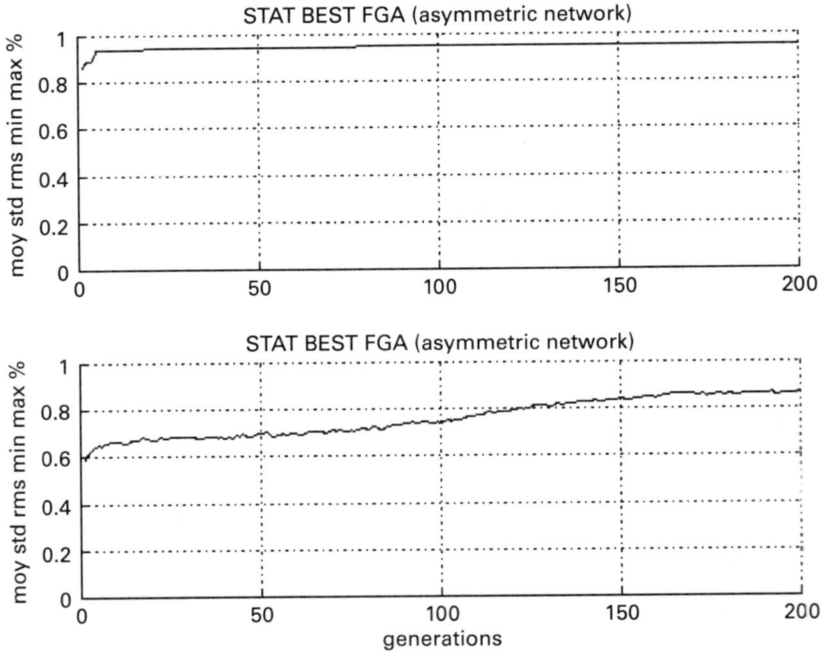

Figure 7-22 FGA stat results for the asymmetric network.

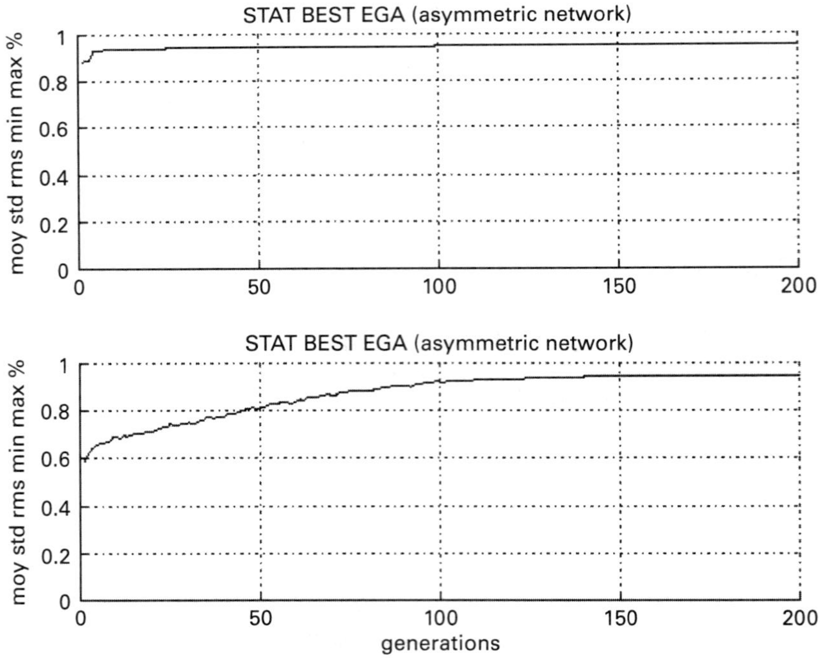

Figure 7-23 EGA stat results for the asymmetric network.

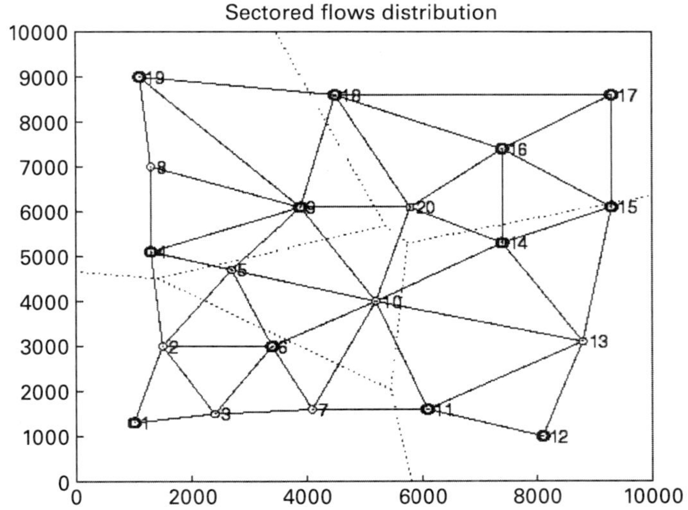

Figure 7-24 Physical sectoring result.

Discrete Approach

Instead of sectorizing the physical airspace this new technique partitions the underlying network connecting the domain points where workload has been registered. The classical graph partitioning problem is usually defined the following way:

> Let $G = (V, E, w)$ be an undirected connected graph, where $V = \{v_1, v_2, ..., v_n\}$ is the set of nodes, $E \in V \times V$ is the set of edges and $w : E \rightarrow N$ defines the weight of the edges. The graph partitioning problem is to divide the graph into K connected components $P_1 \ldots P_k$, such that the sum of the weights of edges between the component is minimal, and the weights of the components are nearly equal.

The problem we have to solve is more complex because when an edge joining two components is cut, a new weight appears on it that is shared by the two associated components. This new weight depends on the cutting flow on the edge and is summarized by the coordination. This means that the global weights of the network before and after partitioning are not the same (the later being heavier). So, the balance for the a priori partitioning (before cutting the edges) is not the same as the balance for the a posteriori partitioning (after cutting the edge).

Constraint Satisfaction

- *Route convexity constraint*: each sector being synthesized by a connected component of a network build from the routes actually used (assumed to be the minimum distance route in the network) this constraint is naturally satisfied.

- *Safety constraint and minimum stay time constraint*: these constraints are partially satisfied by construction of a network where all the small links have been removed. For the longer links (link in the new network), the possible cuts are forced to be in a central zone of the edges, a 7-minute flight distance from both extremities.

Discrete graph partitioning is a classical NP-hard problem (Garey & Johnson, 1979) and then no polynomial algorithm has been identified to solve it. So, the K graph partitioning is usually investigated by iteratively applying a bipartitioning heuristic on the successive created subcomponents (Cheng, 1992; Hendrickson & Leland, 1995), which is definitively a suboptimal approach. Even, for the bipartitioning problem, the most powerful heuristic developed by Goemans and Williamson (1994) is ensured to be at least at 80 percent from the optimum.

So, according to the size of our network (about 1000 nodes), classical combinatorial optimization is not relevant and stochastic optimization seems to be more suitable.

Application to Our Problem

Before an EA can be run, a suitable coding for the problem must be devised.

Data Coding

In our problem, the state variables (which contain all the information needed to define the connected components) consist of the partition of the set of nodes into subsets, each node belonging to one and only one subset (with no empty subset). Furthermore, a power factor is associated with each subset to define the limit between two different components (see figure 7-25).

To create an initial population of individuals (random graph partitioning), the following steps are applied (see figure 7-26):

1. K different nodes are randomly selected from the network [K represents the number of connected components (2 in the example)]. Those nodes are the initial K connected components and are then labeled with different symbols (A and B here).
2. The neighbors of a component $i (i = 1 \ldots K)$ are checked (a node is said to be a neighbor of a connected component if there is a link between this node and a node belonging to the connected component). If this node is free, then it is associated with the component i else this node is already associated with another component and the link joining the current component to the neighbor node is randomly cut into two segments.
3. Step 2 is repeated until all the nodes of the network are labeled.

Recombination Operators

When a population has been created and the best individuals selected, the associated diversity is decreased. To improve this diversity and to have a chance to explore new regions of the state domain, different recombination operators are randomly applied. These operators are stochastic, and modify, more or less, the

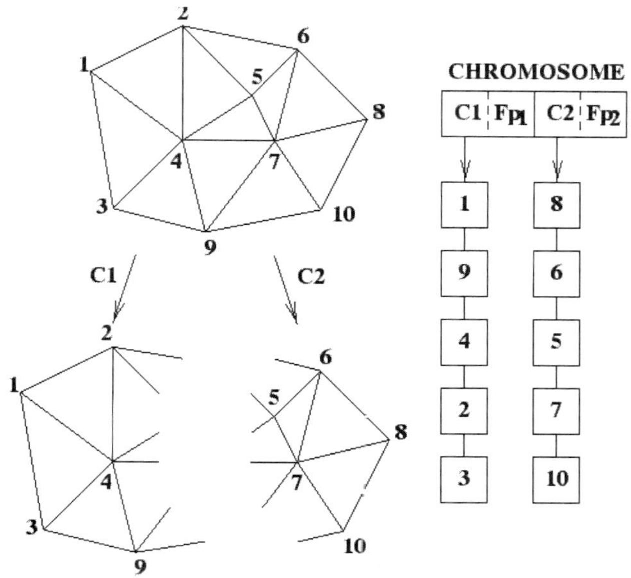

Figure 7-25 Coding of the chromosome.

structure of the chromosome. In our application, cross-over and mutation operators have been developed, but only mutation has been applied because offspring generated by cross-over needed repairing in order to respect constraints. The repairing operator was very penalizing and has been abandoned.

Mutation operators can be classified into three categories (see figure 7-27):

1. *Strong mutation.* This operator modifies all the connected components by randomly choosing *K* new initial nodes and propagating the new components as for the initialization process.

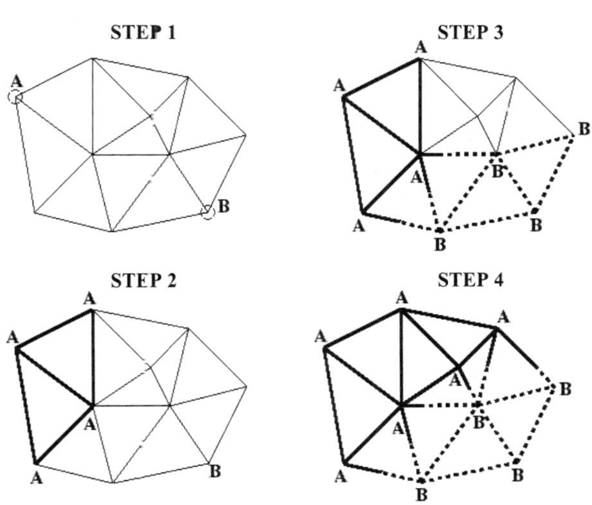

Figure 7-26 Random graph partitioning.

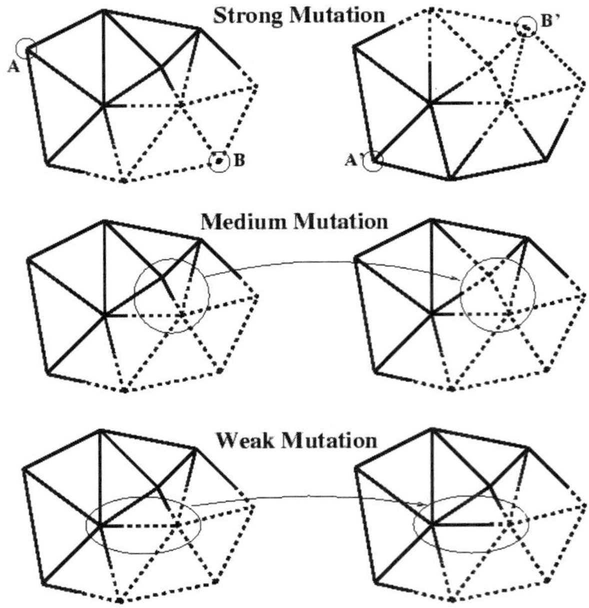

Figure 7-27 Mutation operators.

2. *Medium mutation.* This mutation consists of statistically selecting the most unbalanced component in the chromosome and (statistically) identifying the neighboring component that will better correct the weight of the current component by exchanging one node. After applying this operator one must check that the component that loses one node is still connected. As a matter of fact, this operator can break the connectivity, as shown in figure 7-28. Then, this mutation will be accepted only if the new associated components are not empty and are still connected. To check this last property, a connectedness algorithm is applied to the suspected component. The basic step of this algorithm is the fusion of adjacent vertices. We start with some vertex in the graph and fuse all vertices that are adjacent to it. Then we take the fused vertex and again fuse with it all those vertices that are adjacent to it now. This process of fusion is repeated until no more vertices can be fused. This indicates that a connected component has been "fused" to a single vertex. If this exhausts every vertex in the initial graph, the graph is connected.
3. *Weak mutation.* This operator works the same way as the previous one but only the respective power factors are modified on the two selected components (no repairing is needed).

When a medium or a weak mutation has been decided, the stochastic balancing process induced by there application is applied several times in order to speed up the convergence on the first part of the criterion (balancing).

Application to Test Networks

This method has been successively applied to different kinds of networks with several hundred nodes. To investigate the performance of the algorithm, a series of networks with exact solutions has been used. In all cases, the expected exact

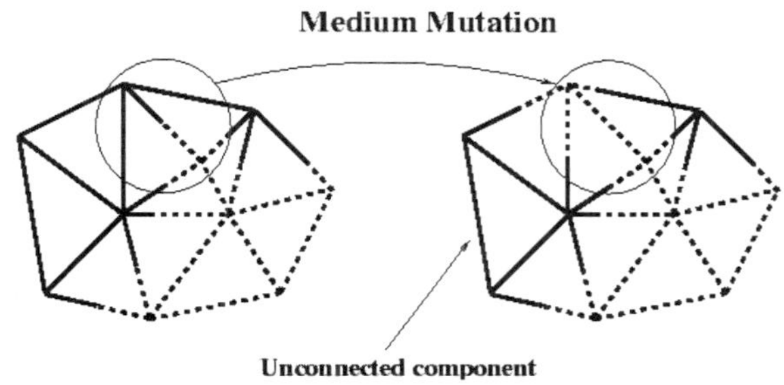

Figure 7-28 Connectivity broken.

solutions have been found but sometimes further exact solutions have been discovered by the sharing mechanisms. Figures 7-29 and 7-30 present two test networks with exact solutions (324 nodes and 400 nodes respectively).

In the first one, an exact solution with 81 components can be identified (this symmetrical solution is trivial for a human because of our brain's ability to investigate symmetries); in the second one an exact solution with 100 components has been hidden in this random network. From the computer, both networks represent the same difficulty but for a human the first one is much easier. The associated EA parameters were the following:

- population size: 100
- number of generations: 100
- probability of mutation: 0.7

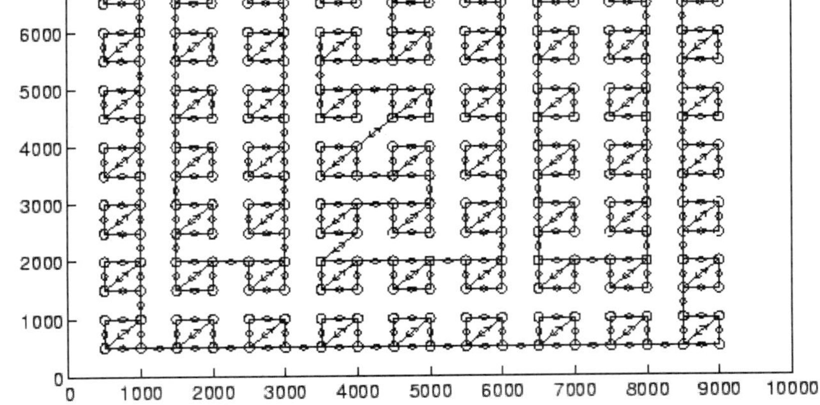

Figure 7-29 Symmetrical test network.

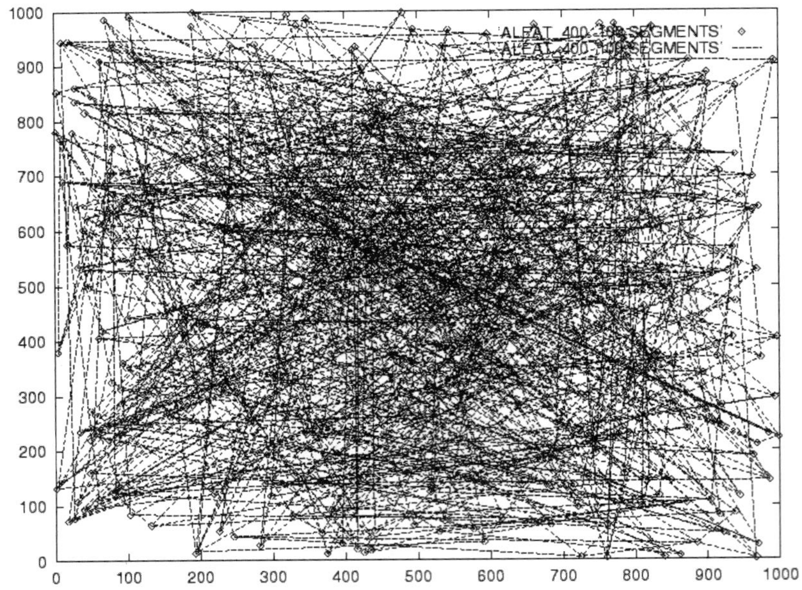

Figure 7-30 Random test network.

- sharing (an adaptative clustered sharing has been used): yes
- elitism: yes

The associated fitness evolutions (best and average fitness on population for each generation) are given in figure 7-31 for the symmetrical network and in figure 7-32 for the random network. In both cases, the fitness reaches 1, which is the optimum according to the fitness calculation (for this network, flows on links joining two square blocks have been set to 0). The large evolution steps are due to the random balancing process during the medium and weak mutations. The execution time was about 10 minutes for both networks on a Pentium (133 MHz).

Conclusion

This study has given a good example of the strong power of evolutionary computation to solve large instances of the graph partitioning problem with very special constraints.

An airspace model has been developed and simplified in order to produce a mathematical formulation that is well adapted to stochastic optimization.

Two kinds of coding have been tried on this sectoring problem. First, a continuous partitioning of the whole airspace with polygonal sectors has been investigated by the use of Voronoi diagrams. To reach this aim we had to extend the chromosome concept to floating strings so that operators do not break the chromosome structure roughly. This modification really improved the algorithm performances regarding the resolution speed and the accuracy of the

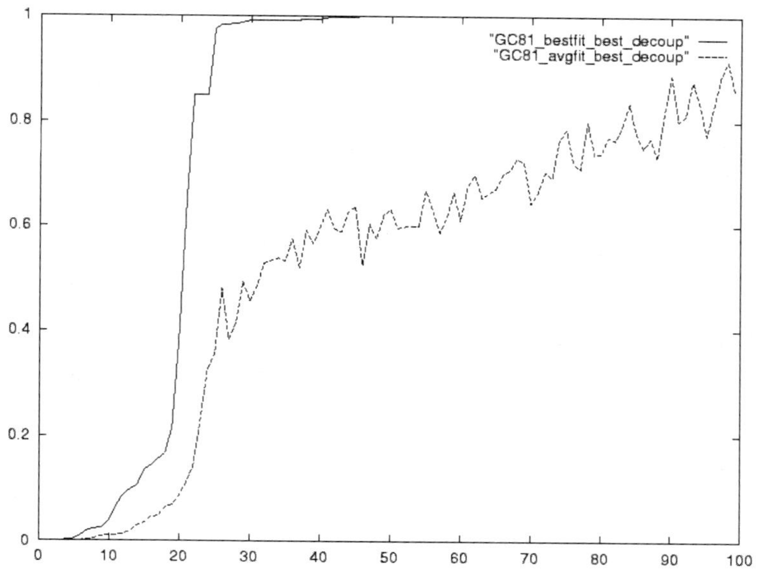

Figure 7-31 Results for the symmetrical network.

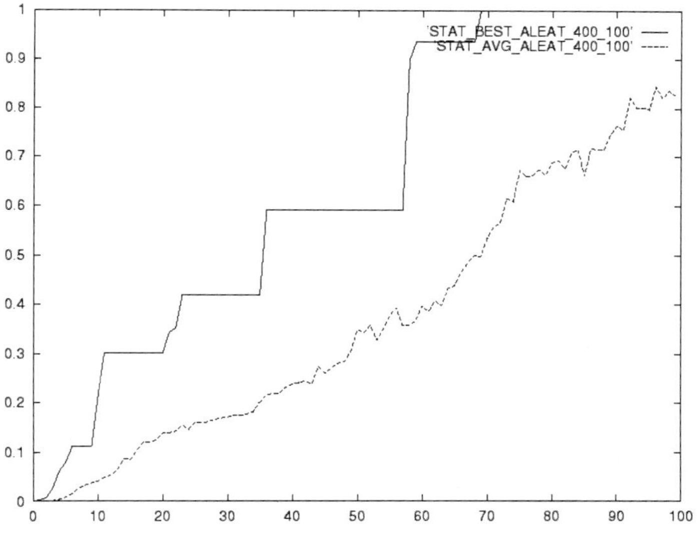

Figure 7-32 Results for the random network.

result. Subsequently a tournament operator has been added, which uses dynamic parameters to improve the space exploration as well as the selection process. This last change brought good improvements to the algorithm convergence rate. As with every genetic algorithm, the key to success lies in the modeling and the operators. Both must be as close as possible to the application problem. In our case, the representation seems to be very close to the physical application, but operators can still be improved, though the ones we have used gave very good results. The strong point of this continuous approach is its ability to partition airspace even if aircraft move out of the regular network of airways. If aircraft stay on this kind of network, a discrete approach can be used, which gives the same quality of results in a shorter time (the speed-up factor is about 50).

This new approach has given very good results and seems to be very well adapted to the air network partitioning problem. It respects the major operational constraints [the synthesized sectors met the route convexity constraint (not the space convexity), the safety constraint, and the minimum stay time constraint] without relaxation (relaxation was used for continuous coding, but then slows down the convergence of the algorithm).

The method used to create the initial network depends only on the traffic itself and can investigate organized traffic or free route (in a free flight environment, where aircraft follow direct routes between origins and destinations without staying on the air network) traffic.

Notes

1. Semicircular rule: aircraft with headings between 0° and 180° have to fly with odd altitude (in hundreds of feet) and even altitude for headings between 180° and 360°.
2. This kind of convexity is stronger than the one imposed by our problem (our sectors have to be convex according to the direction of the links of the network and not in all directions) but is easier to implement.

References

Aarts, E. & J. Korst. 1989. Simulated annealing and Boltzmann machines. New York: Wiley.

Back, T. 1996. Evolutionary Algorithms in Theory and Practice. Oxford University Press.

Cheng, C.-K. 1992. The optimal partitioning of networks. Networks, 22, 297–315.

Delahaye, D. & A. Odoni. 1997. Airspace congestion smoothing by stochastic optimization. In: Proceedings of the Sixth International Conference on Evolutionary Programming. Natural Selection Inc.

Delahaye, D., J.-M. Alliot, M. Schoenauer, & J.-L. Farges. 1994a. Genetic algorithms for air traffic assignment. In: Proceedings of the European Conference on Artificial Intelligence. ECAI.

Delahaye, D., J.-M. Alliot, M. Schoenauer, & J.-L. Farges. 1994b. Genetic algorithms for partitioning airspace. In: Proceedings of the Tenth IEEE Conference on Artificial Intelligence Application. IEEE.

Delahaye, D., J.-M. Alliot, M. Schoenauer, & J.-L. Farges. 1995. Genetic algorithms automatic regrouping of air traffic sectors. In: Proceedings of the Fourth International Conference on Evolutionary Programming. Natural Selection Inc.

Delahaye, D., M. Schoenauer, & J.-M. Alliot. 1998. Airspace sectoring by evolutionary computation. In: Proceedings of the 1998 IEEE International Conference on Evolutionary Computation. IEEE.

Fogel, D.B. 1994. Evolutionary Computation. Toward a New Philosophy of Machine Intelligence. IEEE Press.

Fogel, L.J., A.J. Owens, & M.J. Walsh. 1966. Artificial Intelligence Through Simulated Evolution. Wiley and Sons.

Garey, M.R. & D.S. Johnson. 1979. Computers and Intractability. A Guide to the Theory of NP_Completeness. W.H. Freeman and Co.

Goemans, M.X. & D.P. Williamson. 1994. Improved approximation algorithms for maximum cut and satisfiability problems using semidefinite programming. In: Proceedings of the 26th Symposium on the Theory of Computing.

Goldberg, D.E. 1989. Genetic Algorithms in Search, Optimization and Machine Learning. Addison Wesley.

Hendrickson, B. & R. Leland. 1995. An improved spectral graph partitioning algorithm for mapping parallel computations. SIAM J. Sci. Comput., 16, 452–469.

Ingber, L. & B. Rosen. 1992. Genetic algorithm and very fast simulated reannealing: a comparison. Mathematical and Computer Modeling, 16 (1), 87–100.

Klingman, D. & J.-M. Mulvey (Eds.). 1981. Networks Models and Associated Applications. North-Holland.

Michalewicz, Z. 1992. Genetic Algorithms + Data Structures = Evolution Programs. Springer-Verlag.

Preparata, F.P. & M.-I. Shamos. 1985. Computational Geometry. An Introduction. Springer-Verlag.

Schwefel, H.P. 1995. Evolution and Optimum Seeking. New York, Wiley.

Smith, R.E., D.E. Goldberg, & J.A. Earickson. 1991. SGA-C: A language implementation of a simple genetic algorithm. TCGA report No. 91002, May.

Tuan, P.-L., H.S. Procter, & G.-J. Couluris. 1976. Advanced productivity analysis methods for air traffic control operations. FAA Report RD-76-164, Stanford Research Institute, Menlo Park, Calif., December.

Index